RADIO Q&A
By Bob Grove W8JHD
Founder and publisher of *Monitoring Times*

This collection of readers' questions and their answers from the popular Ask Bob column was selected and compiled from the most recent ten years of publication.

All contents are protected under U.S. copyright law

Copyright 2017

CONTENTS

RADIOS..**P. 3**
FREQUENCIES..**P. 25**
ANTENNAS, GROUNDS, PROPAGATION........**P. 34**
INTEFERENCE...**P. 88**
MISCELLANEOUS..**P. 95**
FREQUENCY ALLOCATIONS TABLE............**P. 125**

Section 1: RADIOS

Q. Upon looking up some receiver specifications, I've come across several abbreviations for units with which I'm not familiar. Could you explain what they are?
A. This broadside list will provide some understanding of those specs. Absolute definitions can be found on the web.

S/N+N: Signal to noise ratio plus noise level is a received signal strength measurement made by starting with the signal to noise ratio, then adding the noise level.

SINAD: Signal plus noise and distortion is another way of expressing the received signal strength.

dB: The decibel (hundredths of a bel) is a relative value comparing signal level to a reference level.

uV: A microvolt (millionth of a volt) is a unit used to express the received signal voltage in absolute terms.

BW: Bandwidth is the actual width of radio spectrum occupied by a signal; it may be anywhere from nearly zero for an unmodulated carrier wave to hundreds of kilohertz for more complex, wideband signals like wideband FM and multiplex.

MDS: the minimum detectable signal is just what it implies, the weakest signal voltage that can still be copied above the background noise.

IP2, IP3: The second and third intercept points refer to the levels at which strong signals begin to distort the receiver's linear amplification, thus producing interfering images (phantom signals). IP3 is commonly called intermodulation ("intermod").

dBm: Decibels per milliwatt is a way to express the power ratio of an amplified signal against a one milliwatt standard reference.

Q. What do TRBO and NFM modes mean, and can they be picked up on any type scanners? Our taxi company has nothing but funny interference noises on their licensed frequencies now.

A. TRBO (MOTOTRBO) is a proprietary digital radio system used by Motorola. NFM is simply narrowband frequency modulation, 6.25 or 12.5 kHz rather than the former 25 kHz, so that more licensees can be crowded into the same amount of spectrum.

Q. Do shortwave receivers with USB/LSB reception have significant advantages over receivers with only AM reception?
A. Upper and lower sidebands are the message-carrying portions of a radio signal. The information is duplicate, so one sideband can be eliminated, thus narrowing the signal bandwidth. The receiver can use narrower filters to extract the signal, thus avoiding the noise and interference experienced with a wider signal.

Q. What is the proper and most effective use of the RF gain control on receivers?
A. The main reason is to reduce signal strengths of extremely strong signals that cause intermodulation interference and "splattering" of adjacent signals. The RF gain control is reduced to the point at which the desired signal is unimpaired by the interference.

Q. I just bought an item advertised as "reconditioned." Does that have a standard meaning?
A. Similar terms include "refurbished" and "remanufactured." While there may be minor cosmetic blemishes, the implication is that it has been tested and repaired if needed to meet original factory specifications.

Q. With the FCC's narrowbanding mandates, what happens to the junked VHF and UHF transceivers?
A. Typically, large lots of surplus radios are offered at auction to the highest bidder. The radios may be used by unaffected services like amateur radio, salvaged for repair parts, or even exported to non-FCC-regulated countries.

Q. Is there any point to owning a desktop receiver if I already have a multiband portable radio?
A. The advantages of a desktop (communications) receiver include the ability to separate closely-spaced signals (selectivity), reject strong-signal overload (dynamic range), reduce phantom signals (images), attenuate electrical interference (noise limiting), using an external antenna instead of the whip, and fine-tune single sideband (SSB) modes.

Q. While driving through the suburbs, I see a lot of abandoned C-band satellite dish antennas, the old "big ugly dishes" ("BUDs"). Are there still services operating on C band?
A. Most C-band signals are digital, and many of those are Free-to-Air (FTA), unencrypted, using MPEG2 or MPEG4 technology. You'll need a dish of 6 feet or larger in diameter, an FTA receiver, and a new, C-band, low-noise, block down-converter feedhorn (LNBF) for successful viewing.

Q. I occasionally will receive subchannels on my digital TV with the message, "Signal Cannot Be Decoded." Are these data channels?
A. It simply means that the signal is too weak or too distorted from multipath to provide a reliable picture. In the old analog days, you would have seen a snowy or blurry picture.

Q. Why do some receiver manufacturers attenuate sensitivity on the AM broadcast band rather than letting the operator choose the option manually?
A. Modern communications receivers have more sensitivity than they need on those frequencies considering the size of antennas typically used with them. The attenuation is done to increase the receiver's immunity to strong signal overload in areas where there are strong broadcast stations.

While that would seem to reduce weak signal reception, it doesn't. At those frequencies, the signal-to-noise ratio is set by atmospheric noise which we hear as heightened background hiss when we connect the antenna. The weak signals do become weaker, but so does the hiss proportionately, so the net result is the signals are heard just as well, and the receiver operates within the limits of its dynamic range.

Q. Why don't radio manufacturers make stand-alone DRM (Digital Radio Mondiale) receivers?
A. It's been tried, but with very limited success. One model, the UniWave, has been discontinued. Very few shortwave broadcasters use DRM and, since it's a licensed system, it raises the cost of the radio. Originally hailed as a salvation for shortwave broadcasters, it hasn't lived up to its expectations.

Q. At the height of the cold war, a red "hotline" phone tied the President's phone directly to the Kremlin. Do you suppose that phone still sits there?
A. Nope. By 1986 the old red phone was replaced by a satellite link for fax/computer intercommunications between the U.S. and Russia.

Q. I hear a continuous siren transmission on one part of the shortwave spectrum. What would this be?
A. If you are listening to an international broadcasting band, it's very likely a deliberate jammer. Political rivals do this to prevent their own citizens from hearing propaganda from other countries. It might get them to think for themselves, and you know how bad that would be!

Q. Do computer-connected shortwave receivers work well in an indoor electrical environment?
A. They work as well as any other shortwave radio fed by an outdoor antenna. If you have an indoor antenna, you will suffer the consequences of electrical appliance interference from just about anything that has electronics in it – especially nearby switching power supplies ("wall warts") so commonly found now instead of the noise-free, transformer-type power supplies.

Q. What are some inexpensive ways to keep a desktop scanner running in the event of a power outage?
A. If it's AC-only powered, you will need an uninterruptible power supply (UPS) like those used for computer backup supplies. If you can operate it from 12 volts DC, the here ae some hints.
1. Select a model with the lowest current rating.
2. Lock out all the channels that aren't vital for reception during battery operation to avoid the extra current required for unnecessary audio.
3. Keep the volume only as loud as necessary.
4. You can run a temporary DC power cord to your car battery or cigarette lighter jack.
5. Keep a good size, 12 VDC, rechargeable battery on hand fully charged.
6. Consider a solar panel that can supply the needed current and voltage to run the scanner during the day and charge the batteries for nighttime use.

Q. The "premium" category of shortwave portable radios has no qualifying members. What do you think of a ham transceiver as an alternative?
A. HF transceivers are specifically designed for competitive reception under noisy, crowded band conditions. No portable ever made has the combination of dynamic range, selectivity, stability, rugged construction, noise blanking, and signal processing available in the majority of amateur HF transceivers.

Q. The term five by five always meant loud and clear to me. Is this an accurate description?
A. "Five by five" is the military equivalent of the amateur "five by nine," the phone version of the Ham's Morse code RST (readability, signal strength, tone quality) 599. For military voice operators, signal strength and clarity are rated one to five. For amateur Morse code, signal strength and tone quality (RST) of a Morse signal are rated one to five and one to nine as indicated above. Thus, five by five, and five,

nine, nine both mean "loud and clear" ("Lima Charlie" in military voice).

Q. I have heard that there are scanners which can detect the presence of old graves. I am a member of the Association of Gravestone Studies. Any idea as to how they work?
A. The real name for these devices is ground-penetrating radar (GPR). They are "scanners" only in the sense that they spread out a beam which scans several feet deep as they are pushed along on their wheels. Reflected signals are traced line-by-line on a video screen to show different soil densities, thus revealing buried shapes.

Q. How can I protect my multiband portable radio from damage caused by a severe electromagnetic pulse?
A. Because of concerns about EMP, extensive tests were performed by a government-approved laboratory to see just what those vulnerabilities were. They built a massive electrical-discharge machine and put a variety of electronic products nearby, some with shielding, some with external wires, some just bare. The results showed that concern for damage from EMP had been greatly exaggerated.

The most vulnerable – and that wasn't many of them – would be electronic devices with long leads on them that would behave like pickup antennas. Most devices without the long wires were pretty safe, especially in metal cases. And all devices were completely safe if enclosed in a simple metal shield.

In other words, if you have a metal box to put the radio in, or even wrap it in aluminum foil, it would be completely protected.

Q. I operate 40 meter CW and our net is often interrupted by a "swish" sound as a carrier glides through our frequency. This is repeated periodically. What is this signal?
A. Ionosondes (frequency sweepers) originate from government agencies (or their contractors) for the most part – NOAA, USN, etc. – and are most commonly sent in pairs or triplets, but certainly can be singles as well.

They are used to determine maximum usable frequencies (MUFs) to intercommunicate among various stations on the earth's surface. They can travel over most any width of spectrum, from a few kilohertz to several megahertz. A spectrum display reveals their moving traces in contrast to frequency-stable signals.

Q. When I hook an outdoor discone to my hand-held scanner, it has poor reception, but the antenna works fine when connected to my desktop scanner. What might be the problem?
A. Since most scanners have virtually identical reception, then:
1. The adapter you're using with the hand-held is not making a good connection, either because of misalignment of, or damage to, the center conductor.
2. Some local signal (FM broadcaster or NOAA weather?) is so strong that it's overloading the hand-held, causing desensitization.
3. The hand-held has lost sensitivity; compare the two scanners with their attachable whips to see if there's a significant difference in weak-signal reception.

Q. I recently purchased a communications receiver, but the manual doesn't explain when I should select such features as filter bandwidth, passband tuning, or noise blanking. How do I know when to select these options?
A. Wide and narrow refer to the bandwidth that the receiver will listen to on a specific center frequency. For example, a typical AM broadcaster may take up 10 kHz of spectrum space, and if you want to hear it crisply, you would select a wide filter like the 15 kHz. If, however, there is another station on a nearby frequency causing interference, you would select a narrower filter to reject that interference while accepting the desired station. In this case, 6 kHz.

Single sideband (SSB) stations mostly use upper sideband (USB), although some lower sideband (LSB) is found, notably by hams below 10 MHz. All require less than 3 kHz of bandwidth, so you'd select the narrower filter (2.4 kHz).

Passband tuning (PBT) is a method of electrically separating an adjacent, interfering signal from one you want to hear. If passband tuning doesn't effectively remove the interference, then you will have to use the narrower filter (The wider filter provides crisper sound quality). In some severe-interference cases you might have to use both.

The pulse noise blanker is useful in a mobile installation for rejecting the "pop-pop" ignition noise, and in the home, office, or other indoor location which is subjected to AC power line noise, fluorescent light noise, and other sources of electrical interference.

Q. I have my shortwave portable radio audio feeding my laptop computer to make recordings, but they are badly distorted. Both are battery operated. What could be the problem?
A. You shouldn't have such a problem if you are using well-shielded cable, stereo plugs on each end, reasonably close impedance matching, and the audio output from the radio isn't overdriving the input on your computer—the most likely cause.

Q. What characteristics define a "superheterodyne" receiver?
A. Edwin Armstrong invented the heterodyne, the mixing of the incoming radio frequency with an oscillator to convert the original signal into two products, one higher and one lower in frequency (the sum and the difference of the two frequencies). The lower frequency was chosen for detection since it was much easier to design.

This frequency conversion scheme was an improvement over the earlier neutrodyne, or tuned radio frequency (TRF) receiver, a series of RF amplification stages followed by a detector.

Armstrong's superheterodyne upgrade added amplification in the successive stages (intermediate frequency or IF). Not long after that, an RF amplifier was added ahead of the mixer for greater sensitivity.

A later variation of the superheterodyne was the autodyne, which employed a pentode (five element) tube which could oscillate as well as convert the incoming signal in the same stage.

Q. I would like to sell my scanner, but it has cellular-frequency coverage because it was made before the ban on cellular-capable scanners. Does that mean I can't sell it?
A. Any product originally certified by the FCC can continue to be owned and resold. Advertise and sell it with a clear conscience.

Q. I'm new to scanning; what are trunking and P-25 and how do I know if I need their capability in my area?
A. If you're in a small town with few communications, you probably don't need them. If you're in a major city, or planning to listen to a nearby major city, you probably do.

Trunking is simply a small pool of frequencies, most often in the 800 MHz range, which are shared by a large number of radios. If each department or division had its own frequency and rarely used it, this would be a waste of valuable spectrum.

In trunking, all of the agencies share the same pool of frequencies; each time one of them presses the mike button, one of the presently-unused frequencies is automatically selected.

If you have a conventional (non-trunking) scanner, as it scans through that pool of trunked frequencies during a busy period, you would hear one agency, then another, then another without the two-way continuity of a single agency. With trunking capability, the scanner is able to follow whichever agency you choose as it switches frequencies, allowing you to monitor that series of unbroken transmissions.

You need to ask a local scanner user, a local law enforcement dispatcher or officer, or a scanner dealer about the system used in your area. You can also visit online radioreference.com for frequency information.

APCO P-25 (Association of Police Communications Officers Project 25) is a digital, non-scrmbled, communications technique which sounds like noise a conventional analog scanner. Keep in mind, however, that there are several digital schemes in prevalent use, such as MOTOTRBO, NXDN, DSTAR, EDACS, and LTR.

Q. Recently I purchased a used car which has heavy static on the AM band, so bad that at times it is difficult to hear the station. When the car engine is turned off there is no static. Any suggestions?
A. The places to look are the ignition (spark plug noise), the antenna coax (open or ungrounded shield), the alternator (whine that changes pitch with engine speed), and the fuel pump (constant whine). Be sure that the radio itself is well grounded to vehicle metal at its support point.

Make sure your vehicle is equipped with resistor spark plugs; ignition noise which is being picked up by the antenna. Resistive, shielded, spark-plug leads are also available. Alternator whine can be suppressed by a husky capacitor (several microfarads, AC rated) across the alternator terminals. If it's the fuel pump, connect a 0.1 microfarad capacitor across its terminals.

Noise suppression kits are available online, from auto parts stores, and the J.C. Whitney catalog.

Q. I recently purchased an analog speech-inversion descrambler manufactured by a U.S. company; I was told that only digital transmissions are illegal to monitor, not analog.
A. That's untrue. Any type of descrambler, analog or digital, has been illegal to own, use, possess, advertise, sell or import since passage of the 1986 Electronic Communications Privacy Act.

Q. Some FM broadcast stations don't regularly identify themselves on a regular basis; is this a violation of the FCC Rules and Regulations? Or have they just caught on to the laxity of enforcement?
A. In the U.S., FM broadcasters must identify their call letters and community at the beginning and ending of the broadcasting day, and as close to every hour as practical during a logical programming break. I suspect there is some laxity among some broadcasters, especially if they are long-time licensees and there's little doubt as to who they are. Nonetheless, I'm sure it's a punishable offense if it is confirmed by the FCC.

Q. What frequency range is used by the Sirius radio satellites?
A. S-band allocations in the 2332.5-2345.0 MHz range.

Q. I have heard of MURS radios; what are they and where can I buy them?
A. MURS (Multi-User Radio Service) was introduced by the FCC in 2000, but not with the hoopla that accompanied FRS (Family Radio Service), probably because retailers were already heavily invested in FRS transceivers.

MURS, however, is a superior service. It allows higher power (2 watts), the attachment of external antennas, and operates at a lower frequency (151/154 MHz), all of which translate to greater range. You can find these imported radios with up to 5 channels on eBay and other Internet e-commerce sites.

Q. Is it possible and legal to use a broad-spectrum jammer to defeat a hidden radio-frequency microphone (bug)?
A. It wouldn't violate federal law prohibiting unlawful eavesdropping, but it would violate FCC Rules and Regulations because its wide-spectrum jamming would disrupt legally-licensed services within its frequency range.

If your concern is to compromise an illicit RF listening device, you'd do better with a spectrum analyzer since it shows all frequency activity within a very wide range and allows you to home in on the device. When found, you can disable it or deliver false information to it.

In the meantime, the best solution for holding a private conversation is away from the office or home, and in a noisy crowd.

Q. When I feed signals from my outdoor antenna through a TV splitter into my two PRO-2052 scanners, I get interference between the two scanners which causes one scanner to stop scanning while I'm listening to a signal on the other.
A. Scanner lockup has been a problem since the invention of the scanner. It is caused by the oscillator circuit of one being heard as a signal by the other. This can happen when:

(1) Sitting side by side, the oscillator signal radiates through the plastic scanner case and into the other scanner, or between attached whips..
(2) The squelch setting is too tight (close to the tripping point), over-sensitizing the scanners' signal sensitivity to each another's radiation;
(3) In your case, inadequate splitter isolation in the common antenna line. Try these two tricks:

When the scanners hang up, simply transfer the active frequency to the other scanner.

Enter the active scanner frequency 5 kHz lower, and the other scanner's lockup frequency 5 kHz higher. This will give you a 10 kHz frequency separation which is less likely to lock up the scanners

Q. I did radio checks in the military and often wondered how an SSB signal could travel farther than a full-carrier AM signal?
A. It doesn't. The same amount of radio frequency energy, radiated from the same antenna on a given frequency at a given time, will travel the same distance. What actually happens is at the receiving end.

A single-sideband signal is narrower in bandwidth (occupies less spectrum) than full-carrier AM, so the receiver's narrower SSB filter reduces the amount of competing noise spectrum. The net result is that it can hear weaker (more distant) signals. The effect is more pronounced with even narrower CW (Morse code) signals.

Q. I have an AM radio tuner with two screws on the back marked "ANTENNA." What's the best way to hook this up, and do I need an earth ground?
A. If one of the screw terminals is marked with a "G" (ground), then it should be connected to the shield of the coax that goes to an outdoor antenna; otherwise, you can choose either screw for the center wire and attach the other to the chassis.

An earth ground won't increase signal strength in reception, but it may help reduce electrical noise interference.

Q. I was given an old receiver that has a 50 ohm coaxial, PL-259 connector for a low-impedance antenna, and a connection for a high-impedance wire connection. Which should I use?
A. No subject brings in more questions than antennas. Use a standard shielded (coax) cable (any impedance) like RG-58/U (50 ohms) or TV style RG-6/U or RG/59 U. At shortwave frequencies, coax lengths under 100 feet or so won't matter.

Connect the shielded cable to the 50 ohm coax connector. You don't need a ground at the far end (at the antenna junction), but if you do have a good earth ground pipe there, it could help reduce electrical line noise interference if you find it objectionable.

Impedance matching for receiving shortwave frequencies is not critical. The primary background noise comes from atmospheric sources (global lightning static), and if one antenna brings in signals stronger than another, it usually means that the signal and the background noise are both louder. There's no real improvement in signal above the background noise.

Q. I have a problem with two local AM broadcasters. I hear them not only on their assigned frequencies of 1480 kHz and 1300 kHz, but on multiples of those frequencies. Is my radio defective?
A. The stations may be radiating harmonics or, if they are extremely close, your receivers may be overloaded, producing spurious intermodulation. An external notch filter or bandpass filter between your antenna and your radio should reduce or eliminate the spurs.

Try a much shorter antenna like a piece of wire to reduce signal strengths. If the drop in level on a spur is much greater than the drop on the fundamental signal, the problem is overload.

Q. Why does the U.S. have so few shortwave AM broadcasters, and no longwave broadcasters?
A. As a long-standing member of the International Telecommunications Union (ITU), the United States has no authorization to utilize the 150-300 kHz longwave broadcast band widely used in Europe, and our

shortwave broadcasters must beam their programming outside of the U.S. since it is not recognized as an American domestic service.

Our 540-1700 kHz medium-wave broadcast band has much better propagation characteristics than longwave broadcasting would. Many American shortwave broadcasters site themselves at the continental borders of the continental U.S. in order to beam their foreign-service signals across the mainland to be accidentally heard domestically.

Q. How is it that I can hear AM and FM radio stations when I'm driving through certain tunnels, even though the earth should be shielding them from reception?
A. It's called "leaky coax." Some cable manufacturers offer coaxial transmission line designed
to let some of the signal out through its shielding, as witnessed in the Lincoln and Holland tunnels,
for example.

Q. I seem to recall a simple formula that can be used to calculate the distance an AM broadcaster can be heard. Can you tell me what that is?
A. There are many factors entering into the expected maximum distance an AM broadcast signal will reach including weather, presence of obstacles, frequency, adjacent- and co-channel interference, antenna gain, antenna pattern, ground conductivity, antenna efficiency, transmitting power, time of day or night, season, sensitivity and selectivity of the receiving equipment, and receiving antenna location, height, directivity, polarization, and gain.

At these low frequencies we factor in three types of wave fronts moving from the antenna: ground waves, space waves and sky waves, all of which are subject to natural fluctuations which affect distance.

Q. My old radio has a "scratchy" volume control. Whenever I turn on my gas stove, the sparks from the igniter can be heard as a loud "tick-tick-tick" on the radio which bring back the sound! What are the causes of the scratchy audio, and are there any fixes?

A. Old potentiometers often suffer disintegration of their resistive element; they can also develop film deposits from grease, salt air, tobacco smoke, and even dust on their elements which cause erratic contact, resulting in scratchy audio.

The pop-pop from the igniter is detected by the radio and may cause an abrupt speaker-cone excursion that shakes the tenuous setting of the wiper on the pot, or it may microscopically realign the pot's element particles. If a mere tap on the volume control resets the volume, that's the likely culprit.

There are essentially three fixes:

1. Rotate the control back and forth a dozen or so times. If that doesn't improve the situation, then:

2. Spray a small burst of contact or tuner cleaner (available from Radio Shack), or drops from an electrical contact cleaning liquid into any access hole of the potentiometer, then rotate the shaft as in 1 above. You can even try a drop or so of the cleaner onto the shaft where it goes into the pot, hoping it will wick down to the wiper and element. If that doesn't do it, then:

3. Replace the pot!

Q. I have a ham transceiver with general-coverage reception. I've run a random wire up to the attic, but hear virtually nothing. Is this due to impedance mismatch?

A. A random, 10-30 foot wire antenna should get scads of medium-strength signals. Impedance mismatch is of minimal impact on reception at those frequencies. During the daytime, try listening to the SW broadcasters in the 11.6-12, 15-16, and 17.5-18 MHz bands. At night you should hear plenty in the 5.8-6.2, 7-7.4, and 9-9.5 MHz ranges. If you hear virtually nothing, then consider:

- A break in the antenna line or receiver connector;
- A panel-selected switching requirement (if it exists) for selecting the antenna;
- A defective, front-end, RF transistor in the receiver;

• A shielded antenna location (metal siding, metalized-Mylar insulation, adjacent heating/air conduit, etc.) preventing adequate antenna reception (put a wire outside for the test).

Q. I would like to get into mobile FM broadcasting so that I can advertise my business on my long commute to and from work. Is this legal?
A. No. By its very nature, broadcasting is FCC-restricted to a fixed geographical point and its distinct boundary. This is to assure that other licensees sharing the same frequencies do not suffer interference from a roving transmitter.

You could, of course, apply for a broadcasting license, then feed over a mobile link to the transmitter site.

An unlicensed (F.C.C. Part 15) transmitter can be used anywhere, but they are very low power, thus restricting the useful range, and they must not interfere with licensed services.

Q. Can you define SINPO code as used in logging shortwave stations?
A. SINPO is an acronym for the characteristics of a received signal: signal (strength), interference (from other stations), noise (natural or man-made), propagation (fade), and overall (total merit). Each is given a value, increasing from 1 to 5.

Q. Recently I experimented with my VHF/UHF antenna by attaching it to an HF antenna tuner (transmatch) and feeding it to my shortwave receiver. I was amazed that I actually had signal improvement over my much-longer shortwave antenna. Does this mean I can scrap the wire and use my VHF/UHF antenna for shortwave?
A. A receiving antenna does not have to be resonant to work well. The antenna tuner adjusted the mismatched antenna to the 50 ohm impedance of the receiver so that there was very little loss in the line. It also acted somewhat like a preselector, preventing swamping from off-frequency signals.

Decades ago, the U.S. Coast Guard determined that a shortwave receiving antenna, properly matched, only needs to be about 5 feet in length to hear virtually 100% of the communications signals. In general terms, an antenna needs only to be long enough to capture enough signal voltage to overcome the receiving system's own internally-generated noise. Once you have higher signal than that, you gain nothing. It also helps to have directivity in an antenna to null out unwanted co-channel interference and favor the desirable signal.

Shortwave antennas are long because then they are naturally resonant at impedances which more closely match the nominal 50 ohm input of most receivers.

Q. I have a wide-coverage receiver with TV reception. In what states is it illegal to tune in and observe wireless cameras in operation?
A. To my knowledge, it's perfectly legal in any state to watch any wireless video camera signal that's on the air. Radio privacy laws protect voice and scrambling only.

Q. Is there any way to receive the police data links on a home computer with a scanner attached?
A. No. Mobile Data terminals like Motorola's MODAT utilize a proprietary system, not in the public domain, and rigidly protected by patent rights. As such, it would be illegal for anyone to market such a product.

Q. How can a receiver radiate a signal that is heard in another nearby radio? I've heard of birdies, harmonics, and images, but what are they?
A. All modern receivers have an oscillator; its radiated frequency and its harmonics (whole-number multiples) can be detected on a nearby receiver. If and incoming radio signal is strong enough, it can mix with harmonics of the oscillator to produce multiple images of the desired signal which can be detected at various parts of the tuning range. These are the "birdies."

Q. Why do some signals get louder (stronger) when I touch the outer case of my portable radio?

A. The conductive salts in your body act as an antenna system, enlarging the radio's capture area, The phenomenon is not likely to be observed on radios with a ground wire or coax-fed antennas because the additional wiring is a far better conductor than you are.

Q. I am considering purchasing a WiFi radio to listen to on-line scanners from various locations. Can I use the radio to access these as well as commercial radio stations?
A. All you need is an internet connection to get any of the scanners or broadcast stations that are streaming their feeds. If you want local AM/FM broadcasters that aren't on the net, then get a WiFi model that has a built-in AM/FM radio that has either its own antenna, or has terminals to connect to an external antenna.

Q. Why aren't there any HF walkie-talkies manufactured for the shortwave amateur bands?
A. Low power at HF can be daunting. During WWII, the BC-611 walkie-talkie was in common use for short-range, field operations, but high-power stations cause considerable interference. Antennas would be of enormous length, and tuned, shorter antennas would have very restricted bandwidth. Electrical and atmospheric noise is higher there than at VHF/UHF, and wide-reaching repeaters are in place everywhere for the readily available and inexpensive VHF/UHF hand-helds. And finally, there's no demand for them.

Q. Scanners with cellular frequency coverage are unlawful in the U.S. Now that all cell phones are digital and cannot be heard on modern scanners, could cellular-frequency-coverage scanners be legally owned by U.S. citizens?
A. No. As outmoded as that regulation is, it is still the law. No one in the U.S. other than a government agency, cellular service provider, or technical laboratory requiring such a device, is allowed to own a scanner that is not FCC type accepted, whether or not it includes cellular frequency coverage.

Q. If I had the right antennas and the correct demodulators, could I receive Free to Air (FTA) programs on a computer-hosted receiver?
A. I'm afraid it's not that simple. The MPEG digitized signals are sent down on TV satellite frequencies in Ku band (12 GHz), and proprietary software is required to receive it on your computer. You can subscribe to these services over the Internet, or you could buy the dish, satellite receiver and do it yourself, but you can't kludge together your own system. The bottom line is that it may be free to air, but it's not free to receive!

Q. What SATCOM equipment is required on ocean-going passenger ships and freighters?
A. The Global Maritime Distress and Safety System (GMDSS) requires all passenger ships and large freighters to use both terrestrial HF/VHF radio and satellite communications to respond to marine distress. INMARSAT C is used for data transmissions including the ship's position; INMARSAT B for radiotelephone and TELEX; INMARSAT FLEET 77 high quality voice and data for fleet service; and 406 MHz Emergency Position Indication Radio Beacons (EPIRBs) automatically transmit position data through the COSPAS/SARSAT satellites for Coast Guard rescue.

Q. When we saw fire trucks going by our house today, we got out a very old scanner that we picked up at an auction years ago, but didn't hear anything. Have emergency communications all gone digital?
A. While digital has definitely taken hold of the public safety agencies, it's been in steps that were bad for scanner listeners for a while until the consumer electronics industry caught up. Here's a simplified chronology:
    1. Years ago, everything was analog (like AM/FM radio).
    2. For security purposes, agencies started using analog scramblers like speech inversion until the consumer market started making analog descramblers.

3. The FCC and Congress banned descramblers that were intended to decode privacy of communications, but these devices were still easy to come by or even make at home.

4. Scanner manufacturers introduced digital scramblers that couldn't be duplicated in the workshop, and manufacturers kept the algorithms (codes) confidential.

5. Along came trunking, whereby a given series of transmissions could keep changing channels and analog scanners couldn't follow them.

6. The scanner industry then made trunk-tracking scanners, so we got back to at least step 4.

7. Following communications failures after 9/11, a reliable digital standard was called for by the government for interagency communications: APCO P-25, an open (public) algorithm, not developed for security, was adopted and has been widely implemented.

8. Scanner manufacturers now make P-25 compatible scanners.

Of course, there's still the possibility that your fire department has simply changed frequencies! One easy way to check local frequencies for your area is to go to www.radioreference.com and look up your county.

Q. I'd like to get into ham radio but don't know where to start, I have so many questions.
A. If you can't find a local radio amateur and don't see contact in local newspaper listings, check with your library to see if they know of a nearby radio club. Sometimes public safety agencies have liaison with hams for emergency communications. Perhaps the easiest way to get help is by visiting the website of the American Radio Relay League (ARRL): www.arrl.org.

Now let's take a look at some of your questions:

In layman's terms, what are the various types of ham radios called?

Originally, amateur radio equipment was separated into receivers and transmitters; now they are combined into transceivers since they share common circuitry. The radio bands (swaths of frequencies set aside for ham radio use) alternate throughout the spectrum with other services that need two-way radio: aircraft, maritime, government and military, business, commercial broadcasting, public safety, and more.

If the radio is small enough to be battery powered and carried personally, it's called a handy-talkie or HT ("walkie-talkie" is a holdover from the cumbersome BC-611 of World War II). Mobile transceivers, as the name suggests, are designed for car installation, and base radios are desktop for AC operation.

Transceivers are offered by frequency range: HF (high frequency) transceivers occupy the shortwave amateur bands (between 1.8 and 30 MHz); VHF transceivers from 50-54, 144-148 and 222-225 MHz; UHF from 420-450, 902-928 and 1240-1300 MHz. Some wideband transceivers combine HF, VHF and even UHF into one transceiver.

Morse code (continuous wave or "CW") is allowed on any amateur frequency, although voice and data are far more popular. Voice may be amplitude modulation (AM is decreasing), single-sideband (AM and SSB are found mostly in the HF bands) or frequency modulation (FM is restricted to frequencies above 25 MHz).

Global communication is conducted on the HF bands, generally below about 10 MHz at night and above during the day due to solar influences on the upper atmosphere (ionosphere) which controls the reflection, absorption and refraction of radio waves.

VHF and UHF communications decrease in distance the higher you rise in frequency. The visual horizon is the norm.

Is it legal for unlicensed hobbyists to use transmit-capable radios as long as they only listen?
Yes.

Are there "best buys" for amateur radio equipment?
Several well-known companies, mostly Japanese, manufacture amateur radio transceivers, including Kenwood, Yaesu, Icom, and

Alinco. While hams have favorites, the fact is that all of these companies produce fine equipment.

    Price is directly related to features, since each additional function adds circuitry. Any amateur radio magazine will reveal virtually all the major and most minor manufacturers over just one or two issues of their advertising. Another source of descriptive and pricing information is from dealers; a comprehensive list may be found at:
http://www.cq-amateur-radio.com/cq_hobby_radio_links/cq_hrl_ar_dealer_links.html

    Do people learn all about this through a license study course?
    Yes. You can take a study course through your amateur radio club, on line, or using various study manuals or mail-order courses. Probably the best concerted source of information is the ARRL, mentioned earlier.

Secgtion 2: FREQUENCIES

Q. Where in the shortwave bands do I listen for ham radio operators, and what modes do they use?
A. The shortwave (high frequency) ham bands are: 1.8-2.0, 3.5-4.0, 5.3-5.4, 7.0-7.3, 10.1-10.15, 14.0-14.35, 18.068-18.168, 21.0-21.45, 24.89-24.99, and 28.0-29.7 MHz. Digital modes (Morse code, RTTY and PSK31) occupy the lower part of these bands, and voice (single sideband or SSB) dominate the upper portions. Lower sideband (LSB) will be heard in the 1.8, 3.5, and 7.0 MHz bands and upper sideband (USB) on the other bands.

Q. What does the phrase "DC to daylight" mean in describing receivers.
A. This description is hyperbole, meaning, quite simply, wide frequency coverage. As electromagnetic waves become higher and higher in frequency, they eventually pass radio wavelengths and become light waves.

Q. I have heard that the government purposely makes GPS readings inaccurate to prevent terrorism; is this true?
A. No, it's not. Decades ago, the U.S. government purposely prohibited accuracy on the GPS readings, but now the military GPS downlinks and the public downlinks have virtually identical accuracy – typically a few feet.

Q. If a clothing manufacturer has embedded an RF ID chip in a product, how would the customer find it? What frequencies are used by these chips?
A. Common frequency ranges are on or around 125-148 kHz, 13.56 MHz, 433 MHz, 902-928 MHz and 2.45 GHz. Moisture and metal take their toll on read ranges, which are from a few inches up to about 30 feet, with some custom (non-commercial) units claiming a mile, depending upon the unit, frequency and power.

Passive tags are "illuminated" by the reader to radiate their code; active tags contain a battery and remain powered as long as five years. Their presence can be detected by a receiver, RF detector, or spectrum analyzer.

Q. Can an ignition key with a memory chip be safely stored in a magnetic box without erasing its memory?
A. Yes, it is an RF transponder system. The non-volatile memory chip in the key has a coiled antenna around it to receive a 125 kHz signal from another coil wrapped around the ignition switch when the key is inserted. That pulse provides enough power to enable the key's chip to send a return signal back to the computer to be recognized, thus allowing operation of the vehicle.

Q. Do electrical utility companies still use power lines to communicate with data and voice?
A. No voice any longer, just data for purposes like remote control and meter readings. Line-load telemetry is found nationally on 154.465 MHz; set your scanner on that frequency and every few seconds you should hear a data stream near you.

Q. Whatever became of TV channel 1?
A. During the 1940s, considerably juggling of frequencies and channel numbers transpired for the emergence of the TV industry and accommodation of the 42-50 MHz FM band. Channel 1 actually existed, and early broadcasters were assigned to use it, mostly experimentally.

By 1948, channel 1, which had occupied various 6 MHz swaths of spectrum between 44 and 56 MHz, was permanently deleted from the allocation plan to make room for the growing land mobile service and the six-meter amateur radio band which would be added later, now occupying 30-50 and 50-54 MHz respectively.

Q. I saw the *Area 51: Fact or Fiction?* special on TV that you were featured in. What did you hear on your receiver, and weren't you taking a risk?

A. Communications at the super-secret Air Force base were pretty routine with little attempt to guard their transmissions. Most of it was air-to-air simulated combat, aircraft flight, and perimeter security. There was some minor digital scrambling.

Our production crew remained on public land and we weren't making nuisances of ourselves, so we didn't pose a threat to the security teams, nor they to us, although they stayed vigilant to our presence. All in all, it was great fun and a delightful experience in the desert north of Las Vegas.

Q. I have noticed that two-meter transceivers (144-148 MHz) seem to outperform 70 cm transceivers (420-450 MHz) at the same power levels. Is this a valid observation?
A. Yes, indeed it is. The higher the frequency, the more attenuation the signal will experience in propagation due to moisture, foliage, smog, terrain, etc. Another factor is the length of the antennas. The longer antenna at two meters captures more signal voltage than the shorter 70 cm, assuming the same ratio fractional-wavelength antennas are used (quarter wavelength, 5/8 wavelength, etc.).

Q. Within the next few decades, is it likely that the shortwave spectrum will no longer be desirable or even useful for emerging technology?
A. The basic task is to provide a reliable, wireless medium for propagating information from one place to another. Emerging technology is moving upwards in frequency, but only for short-range applications, with the singular exception of those long-distance repeaters we call satellites.

Still, satellites are vulnerable, and terrestrial links are often required for backup. For instant, wireless, international telecommunications, shortwave is still the best standby solution.

Military networks worldwide still depend upon the 2-30 MHz high frequency spectrum and, if you take a listen to the 20-100 kHz range, you'll find encrypted, digital, military communications in constant operation in the basement band of VLF.

I suspect that the lower frequencies, right up through shortwave, will remain a dependable mode of informational propagation for quite some time to come.

Q. What HF SSB frequencies would a commercial aircraft pilot use in an emergency when he is out of VHF range?
A. Over land in the U.S., only VHF channels are used. But over the vast ocean waters, he would first issue a Mayday or Pan distress call on one of the LDOC SSB frequencies allotted to his air route and one of the internationally delegated, marine distress/emergency channels like 2182, 4125, 6215, 8291, 12290 or 16420 kHz.

He may be asked to switch to a Coast Guard emergency frequency like 5696 or 8984 kHz for rescue coordination. Upon impact, an automatic distress beacon buoy is deployed on 406 MHz (formerly 121.5/243 MHz) with a swept tone for radio direction finding (RDF).

Q. What frequency does WiFi transmit at? Do their harmonics create interference to other bands?
A. WiFi Internet radios share the FCC Part 802.11 wireless computer service between 2.40-2.48 GHz. Their low power, RF filtering, and rapid frequency hopping avoid causing interference to harmonic frequency ranges as well as to other signals in their own range.

Q. What frequencies are used for keyless-entry systems like garage door and vehicle lock systems?
A. The vast majority of them are on in the 300-435 MHz range. Murata, a leading supplier of miniature oscillator resonators for this purpose, supplies them on 304.30, 315.00, 423.22, 433.87, 433.92 and 434.15 MHz. I've had garage-door remotes on 315 and 389.68 MHz.

Q. I saw an antenna atop a stream gauging system at a local recreational area. What frequencies are used for this, and is it for flood control
A. The hydrotelemetry link that you see is part of the IFLOWS (Integrated Flood Observing and Warning System) network run by NOAA (National Oceanic and Atmospheric Administration). They use

frequencies between 169-172 MHz most of the time, and occasionally 406.xxx and 412.xxx MHz ranges. Yes, it's used for flood control, and also sends data to the NWS (National Weather Service).

Q. Are the channels occupied by digital television different from the old analog frequencies?
A. The DTV channels are simply reassigned spectrum taken from the former analog TV channels 2-69. DTV multiplexes several different programs simultaneously; you choose a "major" (conventional) channel number, then a specific program sub-channel, like 2-1, 2-2, 2-3, etc. up to -99. Data channels are numbered -100 to -199.

Q. While listening to air traffic in the 118-137 MHz band, I occasionally hear reference to a "squawk frequency"; what is that and what is it used for?
A. "Squawk" is simply the reference to activating a radar-frequency transponder that will distinguish a particular aircraft on a busy radar screen. If the aircraft pilot responds to "Squawk 7441"; that simply means he will press that series of numerals which will identify which blip is his on the tower radar screen. It can also notify the Collision Avoidance System on other aircraft.

Q. Are cell phone conversations any more secure today than years ago when anyone with a scanner could listen in to the call? Are the same frequencies used now and are conversations encrypted?
A. The same frequencies are still used, and the voices are now digitally encrypted. No receiver, scanner, or decoder can be sold in the U.S. that will allow you to listen to the conversations.

Q. Did narrowbanding require licensees to change their frequencies? Do I have to change frequency entries into my scanner for agencies I already have?
A. No, none of the current licensees have to change frequencies, only the bandwidth of their transmitted signal. Your frequency entries may remain as they are.

Q. I am listening now on 15.016 MHz to what sounds like military communications in Upper Sideband (USB). They start off by saying "Attention all stations," and then go into what seems to be a numbers-and-alfa cadence.
A. You have tuned into the USAF High Frequency Global Communications System (HF-GCS) on which they transmit scheduled Emergency Action Messages (EAMs). Another active frequency is 11175 kHz.

Q. What did the FCC narrowbanding regulations of 2013 actually do, and to what services?
A. The mandate to narrow the channel bandwidths for VHF/UHF FM communications applies only to private land mobile licensees (public safety, business and industrial), both analog and digital, in the 150-174 and 421-512 MHz bands.

Private land mobile transmissions other than paging in the 150-174 and 421-512 MHz ranges now employ no greater than 12.5 kHz bandwidths, rather than the former 25 kHz bandwidth allowance. This made room for two signals in the spaces between former, consecutive, single-signal channels. For example, where two former licensees could occupy adjacent channels such as 460.250 MHz and 460.275 MHz, now another licensee may be sandwiched between them on 460.2625 MHz.

It did not affect amateur radio, CB, GMRS, FRS, MURS, marine radio, aircraft, TV, FM broadcasting, or NOAA weather. Nor did it apply to the 30-50, 72-76, 216-222, 700, 800, or 900 MHz bands. It had no affect on federal government communications since they were already narrowband.

Licensees did not have to change their authorized frequencies, only reduce their bandwidths. Theoretically, this results in a 3 dB reduction in signal strengths. In many cases, the 3 dB lower signal strength wasn't noticed, but where it was, the loss was made up for with higher power, directional antennas, more repeater sites, and taller masts.

Q. How did narrowbanding affect scanner listening?
A. If your old scanner can only program larger channel spacing like 15 kHz intervals, you are likely to lose reception of those new, narrower channels. If it allows narrower spacing like 5 kHz, then even if it's not right smack dab on the center of the channel, it will still pick up the signal due to the wide shape factor of scanner filters.

However, the new, narrower modulation (deviation) means that some models without automatic gain control (AGC) will suffer a quieter audio recovery than when listening to older, wider deviation signals. The fix? Just turn up the volume a bit.

Q. What are some of the common frequencies used by law enforcement video and wireless mikes on police stops?
A. The most common are in the 169-172, 174-218, and 902-928 MHz ranges. Former 698-806 MHz frequencies have been banned by the FCC. Newer units use the 2.4 GHz WiFi band, and some are in the 5.8 GHz microwave band.

Q. I used to be able to hear the wireless intercoms at drive-through restaurants with my scanner, but no longer can find them at my Wendy's even using autosearch function. Could they have gone digital?
A. Yes. There is a trend upward in frequency (1.9 and 2.4 GHz bands) and using digital modulation. A company called Quail Digital specializes in such wireless headsets for the drive-through trade.

Q. What frequencies and modes do the Mars Rovers operate on?
A. When operational, Curiosity transmits around 401 MHz to the transponder (MELACOM) on the Mars Reconnaissance Orbiter (MRO), which then talks back in the 437 MHz Amateur Satellite Service band via two patch antennas. The MRO can talk with any of the Martian Rovers using CCSDS Proximity-1 protocol. A dish antenna transmits to Earth in the 7-8 GHz X-band and 421-512 MHz bands.

Q. Can Earth's shortwave broadcasts be heard from outer space, and have astronauts in orbit ever monitored these broadcasts?

A. Yes, signals in the HF spectrum can penetrate the ionosphere and radiate into space depending on solar influences which change throughout the day and night. The higher the frequency, the more likely they escape. We can listen to Jupiter's electrical storms on 20 MHz, and 15.016 MHz SSB was a backup frequency for the Gemini space program. It is unlikely, however, that the astronauts made any attempts to listen to international broadcasters during their busy missions, nor that they had tunable radios with which to do it.

Q. I heard short bursts of digital modulation on 154.570 and 154.600 MHz. Aren't these both MURS allocations?
A. Indeed they are, right along with 151.820, 151.880, and 151.940 MHz. The Multi-Use Radio Service (MURS) allows unlicensed voice or data for intercommunication or remote control and telemetry. Continuous transmission is prohibited.

Q. Do all cell phone systems use the same 800 MHz frequencies?
A. Not by a long shot. If we consider North American along with European/Asian allocations, currently used bands include 806-960, 1710-1785, 1805-1990, and 2110-2170 MHz in various fractional combinations among licensees and countries.

Q. I recently ran across a very weak signal on 950 kHz. It appears to be a time signal with just a tick every second and a quick Morse code ID. I only receive it at night. Any idea what it is?
A. These are time ticks from Cuba's Radio Reloj (Spanish for "clock"). They are heard on 570, 820, 860, 870, 950, 960, and 1020 kHz.

Q. I have been hearing USB transmissions between 4100 and 4300 kHz. They exchange greetings and signal reports. They sound like amateur radio, but they are outside the ham band. They give their call signs in the format AAA#AA. Am I hearing MARS, marine, or what?
A. MARS (Military Affiliate Radio Service) occupies that and other parts of the spectrum. They were authorized frequencies adjacent to the ham bands so that amateur transceivers could be used for MARS

communications. Even though military communications don't require individual operators to be licensed, MARS does require amateur radio licenses.

Q. Decades ago, police transmissions to patrol cars could be heard just above the AM broadcast band. When did they cease?
A. The 1600-1700 kHz police allocation disappeared by the early 1950s. Since then, the AM broadcast band was extended into that former spectrum.

Q. Why do I receive greater distance on the higher shortwave frequencies than on the lower?
A. At the lower frequencies, signals propagate more directly, often following the curvature of the earth and are absorbed by the atmosphere, while at higher frequencies they take off over the visual horizon and proceed skyward and above the atmosphere where they are more likely to be reflected by the ionosphere back down to the earth at great distances.

Q. I have been given an atomic clock which monitors the 60 kHz frequency standard station in Colorado. At those low frequencies, how can it send enough data to set my clock? Isn't there a time delay for the signal to be accurate over distances?
A. WWVB continuously broadcasts time and frequency signals at 60 kHz at a data rate of 1 bit per second using pulse width modulation. The code contains the year, day of year, hour, minute, second, Daylight Savings Time, leap years, and leap seconds.

VLF signals travel by ground wave; the delay of arrival anywhere in the U.S. is 1/100 second or less. I doubt that your clock can display time that accurately.

Section 3: ANTENNAS, GROUNDS, AND PROPAGATION

Q. I just got a TV dish installed. Why do I lose the signal when rains move in?
A. Signals at microwave frequencies are absorbed by moisture, that's why receiving dish antennas have to be pointed away from intruding, leafy trees, and why rain-saturated clouds can block signals.

Q. Will an antenna originally designed for analog TV or radio reception work for receiving digital or high-definition broadcasts?
A. Absolutely. The modulation scheme of a transmission has nothing to do with the antenna. If the new mode is on the same frequency as it was before, any antenna designed to receive that frequency will receive any mode of transmission on that frequency.

The only difference is that digital signals need to be perfect for any picture at all. In the analog days, a fringe signal could still be seen on a snowy screen, but such a digital signal now would be totally blank or a torn patchwork of distorted pixels at best.

Q. Why were 50 and 75 ohms chosen for coaxial cable impedances?
A. The important considerations for transmission line are the breakdown voltage and loss of the inner insulation, and the power-carrying capacity of the conductor. Since these factors operate independently of one another, no single impedance is a perfect match for all considerations. A compromise between optimum power-carrying capacity (30 ohms) and breakdown voltage (60 ohms) was necessary.

The second reason was that a basic, half-wavelength dipole in a typical terrestrial installation has a center feedpoint impedance of about 50 ohms

During the 1940s, a Radio Manufacturers Association (RMA) committee recommended to the U.S. Navy a standard of 50 ohms, an impedance that could be achieved when using commercially-available copper water tubing for the conductors.

So far as 75 ohm coax for TV and video distribution systems which carry very low signal levels, power-carrying capacity isn't a problem, nor is breakdown voltage. Only the dielectric attenuation contributes significant loss, and that is lowest at 77 ohms impedance.

Q. When you compute a half-wave antenna length by dividing 468 by the operating frequency in MHz, is the resonance of the antenna dependent upon the impedance of the feed line?
A. The resonant frequency of the antenna is independent of the feedline impedance no matter where you feed the antenna along its length. The impedance of the antenna wire is a function of its length, local environment, operating frequency, and where you attach the feed line.
The impedance of the transmission line is determined by its own distributed capacitance and inductance, not what it is attached to.

Q. What is the difference between an antenna tuner and a preselector? Will one or the other reduce electrical noise interference on my shortwave receiver?
A. An antenna tuner (more correctly, a transmatch) is a device which matches the impedance of your radio (typically a transmitter or transceiver) and your antenna system in order to minimize transmission-line losses. Since a given antenna length varies in impedance (a combination of inductance and capacitance) with changes in frequency, the inductance and/or capacitance of the transmatch has to be retuned as well.
A preselector is also a frequency-adjustable combination of inductance and capacitance placed between an antenna and a receiver, but its purpose is to appear passive to the desired signal frequency, but to attenuate off-frequency signals to minimize interference from them. If the noise is on the same frequency as the desired signal, a preselector will not reduce the interference.

Q. I am just getting back into shortwave listening and would appreciate a few hints about antenna installation.

A. Most important is distance from interference generators such as indoor electronic accessories, household wiring, and power lines. It would be best to run the wire away from the house to a tree rather than confine the antenna to the roof. Bring the signal in with coaxial cable to prevent the intrusion of indoor electrical interference on the feedline.

A height of at least 15 feet above ground has always worked well for me. Wire length of some 25 to 50 feet or so is certainly adequate for modern, high-sensitivity receivers. Naval communicators use much shorter antennas and get excellent performance. Since you are going to be tuning your receiver over a 30:1 change in frequency range, there is no specifically-resonant antenna length.

An east/west wire alignment for a typical dipole will receive north/south signals off its sides, favoring Europe and Asia as well as South America and, to a degree, Africa. Lesser heard will be off the ends of the wire.

An actual earth ground is probably unnecessary. You can try a ground rod to see if it helps reduce local noise interference, but it won't increase signal strength.

The use of a gas-discharge lightning arrestor is recommended. Nothing can survive a direct lightning strike, but nearby strokes can induce high voltages on an antenna line that can damage a radio, and the arrestor short-circuits these harmlessly.

Q. I currently have a shortwave wire antenna mounted in my attic. Would I be better off with an outdoor antenna? I'm concerned about tree limbs, and wonder about angling the wire antenna.
A. Outdoor antenna locations are *always* better than indoor. At shortwave, the indoor antenna is subjected to electrical wiring noise, reflective/absorptive effects of wiring and ducting, shielding from metal roofing or metalized Mylar insulation in the walls, microprocessor radiation from electronic accessories, and some other electric/electronic devices as well.

A great outdoor combination consists of two horizontal dipoles at right angles with two coax lead-ins; you can switch between them for either reduction of interference or enhancement of desired signal, depending upon the compass direction.

If you mount a dipole at an angle (known as a sloper), it becomes directional toward the lower end. Tree limbs don't really have much effect on shortwave reception, so erect whatever is convenient. For all-direction response, a vertical is a good possibility. Mount it away from the house and electric wiring to minimize electrical noise pickup.

Q. I would like to mount a scanner antenna in a tree and disguise its elements with paint. Will that affect its performance?
A. The short answer is no, just so long as you don't cover the insulators with paint. Darker colors like black often have conductive materials in them which could effectively short-circuit signals between the elements before they get to the transmission line.

Practically, however, it's doubtful that even with paint on them, the conductance of the paint when dry would be low enough to appreciably reduce signal strengths. But it's better to be safe than sorry.

Q. How do I know if my gas-discharge lightning arrestor has been hit by a surge? Do they need periodic replacement?
A. The gas-discharge lightning protectors I have on my antennas have been there for decades. They defend against voltage induced on an antenna line by nearby strikes, but nothing can survive a direct hit.

If the voltage from a nearby strike was high enough to ionize the gas (electrically charge it so it conducts), it recovers immediately. It can do this nearly indefinitely considering the rarity of such events. If a strike is close enough to damage the device, you would either notice a reduction in signal levels from metal particles shorting out the device, or notice nothing at all because it may have merely released its gas through a crack in the seal, in which case it's no longer protective. If you can see the glass envelope, the appearance of darkness on the glass may indicate vaporized metal in a damaged protector.

It's a good idea to periodically examine the entire antenna system for damage from wind, lightning, and coax-chewing squirrels.

Outdoor coax should be replaced every five years or so, and antennas should be replaced when their joints and contacts exhibit corrosion unless those joints can be cleaned and weather protected.

If you have an ohmmeter, you can remove the protector from the circuit and measure its correct conductivity and insulation: There should be a dead short (0 ohms) between the input and output center pins, and completely open (infinitely high resistance) across either of the connectors from the center to the shell.

Occasionally I tune in a local frequency of known signal strength to test my antenna system, like NOAA weather or commercial AM, FM and TV broadcasters.

Q. I live in an area that has marginal reception for weather radios; sometimes I get the alerts, sometimes I don't. The radio has a jack for an external antenna. Can I simply attach a random wire antenna to it, or would something else be more suitable?

A. A random "longwire" antenna will only work if it just happens to be the right length for an appropriate harmonic of its naturally-resonant frequency, and in the vertical plane. If these conditions aren't met, it will only hear VHF signals if they are so strong that they overcome the poor suitability of the random wire length and configuration.

You can optimize reception with a simple, outdoor, center-fed, 32-inch vertical dipole, fed at the center with low-loss coax like TV-type RG-59/U or RG-6/U. If the signal is reasonably strong and you aren't using more than about 50 feet of cable, standard RG-58/U "CB style" coax should work.

Even better is a "drooping radial" ground-plane antenna; these have been used by hams on VHJF/UHF frequencies for decades. All you need are a female, chassis-mount, SO-239 coax connector and five 16" lengths of stiff copper wire.

One wire is soldered straight up from the center solder pin with the connector sitting flush on its threads. The other four are soldered through the mounting holes, flared downward at roughly a 45 degree angle.

The nice thing about this home-brew antenna is that it can be simply set on a convenient length of pipe. The coax goes up through the pipe and is screwed onto the SO-239 connector which simply sits in the hole atop of the pipe.

A side benefit of this antenna is that it makes a dandy scanner antenna for the 150 and 450 MHz bands, and if you use 19-inch lengths for the elements, it's a great two-meter (144-148 MHz) and 70 cm (420-450 MHz) transmitting antenna.

Q. Can a discone antenna be used as an effective TV antenna to receive distant VHF and UHF channels?
A. In a word, no. While discones can cover the majority of the VHF and UHF TV channels, the broadcasters transmit their signals horizontally polarized so that traditional TV beam-style antennas can be used to receive them. Discones are vertically polarized, so you already start with a substantial signal loss from cross polarization.

If, however, the signals have reflected a great deal over that distance, the polarization is probably mixed enough that it doesn't matter. Even so, discones have no gain, so the beam is still better.

Q. What is the antenna impedance of a long wire? Can I calculate it?
A. A longwire is a general reference to any wire antenna more than one-half wavelength long for a specific design frequency. For tuning considerations, wires longer than a half wave have high impedance at their feedpoints (generally several thousand ohms) and are inductive; wire antennas shorter than a half wave are low impedance and capacitive. Thus, a half-wave dipole for the 40 meter band (7 MHz) is a longwire on 20 meters (14 MHz), and a short, low-impedance antenna on 80 meters (3.5 MHz).

Antenna impedance is a complex value consisting of radiation resistance plus capacitive and inductive reactance. The desire is to cancel the two reactances (opposition to signal flow), leaving just the radiation resistance.

You cancel a capacitive reactance with an inductance (coil), and an inductive reactance with a capacitance (capacitor). In the transmatch ("tuner"), the capacitor is an air variable, and the inductance is either a tapped or rolling-contact coil.

Software and mathematical models are both found in the ARRL Antenna Book from the ARRL Book Store:

http://www.arrl.org/shop/What-s-New

Their software can be trusted.

Q. Recently, a distant FM broadcaster added a repeater near me which is on the same frequency as my favorite FM station. Will an FM beam antenna be sharp enough to separate the two signals? (Robert Compton, email)

A. A directional (Beam) antenna would probably do the job. A great deal depends on the relative strengths of the two signals, and how far they are separated on a compass circle with you at the center. Whether an FM station or any other VHF/UHF communications signal, the antenna consideration would be the same—the higher the gain, the sharper the directivity.

You don't necessarily have to point it toward the desired signal; try rotating the antenna so that the offending signal arrives off the side where there is a deep rejection null. The higher the gain, the sharper and deeper the null. Ideally, of course, the desired signal would be in front of the antenna, and the interfering signal off the side.

Q. Is there a general formula for determining the listening distance for a scanner?

A. Yes, but it's approximate because of the many factors that limit range, such as frequency, weather, terrain, cable losses, obstructions, tower height, transmitter power, antenna locations, antenna gain, and receiver sensitivity.

The visual horizon in miles between two antennas is found by taking the square root of (1.35 x A), where A is the combined height in feet of the two antennas. Since radio waves bend somewhat toward the curve of the earth rather than fly straight-line into space, the actual radio range is considerably greater than this calculation shows.

If an elevated, outdoor, omnidirectional receiving antenna and low-loss cable are used, terrain is reasonably flat, and the transmitter operates at 100 watts or so, your scanner should be able to hear mobiles 15-25 miles away, and base stations 50-75 miles away. Adding a directional beam can increase this to 75-100 miles.

Q. I bought a rooftop scanner antenna, but it's not picking up signals as well as the little indoor 18" whip. I've even replaced the antenna and balun transformer with the same results. What gives?
A. Assuming that the balun transformer and antenna are operating properly, the problem is most certainly in the cable or connectors.

Make sure your cable's center conductor is long enough to insert into the connector onto which it screws. Test the cable with an ohmmeter for shorts and opens. You should get a reading of a few ohms from end to end of the center wire and the same when you measure end to end of the shield, but you should get no reading (infinite resistance) when you measure center wire to shield.

Occasionally, while fitting a connector to a socket, the center wire gets bent over; check that, too.

Q. I have an active antenna for shortwave reception. It works fine with the telescoping whip extended, but when I substitute a similar-length fiberglass whip, reception is not as good. Why not?
A. Is the whip spiral wound? The inductance could play a part in frequency favoring. If it's simply plated, the whip could have a corroded or broken contact.

Q. I see a lot of audio and TV antenna cables being offered with "gold plated" connectors. How much better will such cables perform over nickel or any other plating? Is there enough gold on those connectors to justify the extra expense?
A. The only benefit that gold plating provides is immunity to corrosion which would impede electrical conductivity between the connectors. Gold is actually a poorer conductor than copper, and even if it would be a better conductor, that thin film would do nothing to enhance the signal

transfer because it still has to go through many feet of copper wire. The best conductor is actually silver, but silver plating would invite tarnishing in the weather. The simple answer is that no plating will do anything more than what's provided by clean contacts.

Q. What coax would you recommend for use in 2.4 GHz work?
A. Additional information is needed such as whether it's for transmitting; how much power and at what impedance; the length of the line; types of connectors required and, if cost is an object. In general, stay away from most thin, highly-flexible cables like RG-174 and RG-58. For short runs (a few feet), RG-214, RG-8, RG-59, and RG-6 are all good choices. On longer runs, LMR-400 is recommended and widely available; increasing LMR numbers (500, 600, etc.) indicate even better performance at higher prices.

Q. RG-8/U coaxial cable is designed for communications, yet you recommend RG-6/U for scanners when it's designed for TV. Why?
A. RG-8/U foam-dielectric is excellent, low-loss, coaxial cable. It's also large, heavy, expensive, stiff, requires large connectors, and is designed for transmitting. RG-6/U is just as low-loss, smaller, lighter, more flexible, cheaper, adaptable to a wide array of connectors, and is designed for receiving signals from DC up through at least 1000 MHz.

Q. Have you ever tested one of the little stickers that go behind a cell phone battery that are supposed to boost the signal? Do they work?
A. Yes, I have, and no, they don't. These little printed pieces of paper don't even conduct a signal, and even if they did, they don't stand a ghost of a chance of doing anything behind the battery! They are a scam.

Q. How close can multiple mobile antennas be spaced before they negatively interact with each other?
A. As a rule, to prevent distortion of the omnidirectional pattern of any antenna, keep them at least 1/4 wavelength apart as measured at the lowest operating frequency. However, this applies to antenna elements

of the same resonant frequencies. For example, even though a quarter wavelength for a 27 MHz CB antenna is about 9 feet, you could bring one within a foot or so of a 144 MHz ham whip without serious alteration in the pattern of either antenna because they are not resonant at the same frequency range, nor even harmonically related. But if you are operating two 144 MHz antennas atop the vehicle, separate them by at least 1/4 wavelength (18 inches) to prevent distortion.

Q. I've seen references to a "dipole cluster"; just what does that mean?
A. A dipole is the simplest antenna—a long conductor, cut at its center and attached to a transmission line. For best impedance matching at 6, 11 and 15 MHz, we could combine three dipoles—one 78' long, another 42' long, and the third 31' long-- on one transmission line, each working on its own resonant frequency range. That's a dipole cluster.
    Another technique is to add several graduated lengths, separated by coil inductances, into one long dipole for multiple band coverage.

Q. Can I bring my car scanner antenna indoors and use it with my base antenna?
A. If the mobile antenna is mounted on a large, metal surface, yes, since the car body serves as half the antenna. Many folks simply assume they can bring a magnetic-base mobile antenna inside, set it on a table, and it will work as well as on the car roof. It won't. Affix it to the top surface of a file cabinet, or even invert it and stick it to a ceiling vent that is connected to metal ducting. A massive metal mount is important.

Q. Why should we have to ground a receiver or transceiver when the plug to the unit already has a ground in the third prong of the plug?
A. The third-wire ground on modern electrical appliances is a shock preventive only. With the random lengths of line cords and household electrical wiring, any impedance match between the radio and an earth ground at radio frequencies would be highly unlikely at worst, and accidental at best.
    Think of an RF ground as part of the antenna system, requiring a low standing-wave ratio (SWR) to effectively couple the radio ground

to the earth. House wiring just doesn't do it. Even though the random house wiring might provide the right length at some frequencies, that's pretty iffy!

Q. Exactly where is the North Pole located (please don't say 90 degrees north).
A. The north geographical pole is in the middle of the Arctic Ocean and covered with sea ice. It is the point around which the Earth spins as illustrated on any world globe by the pivot point at the top.
The north magnetic pole is the location at which a compass needle would point downward toward the center of the earth. As of 2017, it was located at 86.5 degrees north, 172.6 degrees west, and was slowly drifting toward Russia.

Q. I have a random wire antenna in my attic connected to two receivers simply wired in parallel at their antenna inputs. Is this the proper way to connect two receivers to one antenna? It seems I'm losing signal strength in one of the receivers.
A. If you have two receivers with identical 50 ohm antenna receptacles, you would theoretically lose only 3 dB (half an S-unit) in each since the available signal is now split in half for each radio. Losing an appreciable amount of signal level in just one radio is wrong. The two inputs are probably designed for different impedances.
     Rather than hardwire the two receiver inputs together, it's best to use a splitter to isolate them. Most standard TV antenna splitters work fine at shortwave frequencies. If your split signal levels are too low, you could add an in-line preamplifier.

Q. I have been unable to find any information on how tapering a whip antenna toward the top affects its resonance or signal. What is the relationship of the length and diameter of an antenna element?
A. The gradual tapering of a whip antenna has little effect on its electrical characteristics, it's only a means of providing flexibility in mobile applications where it's likely to strike something overhead.

In vertical HF base antennas, it simply makes more sense from a weight/leverage perspective to make the telescoping sections gradually smaller toward the top than at the bottom. Base VHF and UHF antenna elements are rarely tapered.

Q. With the glut of old rooftop TV antennas available, can these piles of scrap be rolled over 90 degrees so they are vertically polarized and used as beam antennas for scanner reception?
A. Absolutely. The legend Grove Scanner Beam was based on just such a TV antenna design with the elements cut slightly different in length and spacing to favor the land mobile bands, but even a TV antenna does a good job on scanners.

Q. I have a collection of aluminum tubing from scrapped antennas. How can I calculate the lengths I need for scanner listening?
A. If you're going to leave the elements paired as a ahlf-wave dipole, the basic formula for calculating the total length in inches is to divide 5616 by the frequency in MHz. Therefore, you would use:|
    135 inches (67.5 inches each side of the insulator) for the 30-50 MHz low band;
    36-inches (18 inches each side of the insulator) for the 144-174 MHz high band;
    14 inches (7 inches each side) for 406-512 MHz UHF; and
    8 inches (4 inches + 4 inches) for the 806-960 MHz band.
    It really doesn't matter whether the coax shield or center conductor is connected to the upper or lower element, but most people feel better when the "hot" center wire is connected to the upper element. Mount the dipole vertically to match the polarization of local signals, and bring the coax out from being directly alongside the lower element.

Q. I've often heard that antenna tuners are needed on a portable shortwave radio, but you say that it's necessary for sending and NOT necessary for receiving. Why the difference of opinions?
A. Much of the difference in opinion comes from listeners' different expectations and experiences with different radios. Portable, multiband

radios are far more vulnerable to overload problems than desktop communications receivers and amateur radio transceivers.

If you are in a metropolitan area, and your shortwave reception is compromised by strong-signal overload from nearby transmitters, then a frequency-adjustable preselector, not a tuner, would solve the problem. A preselector isolates a narrow swath of spectrum, deeply suppressing frequencies above and below that.

An antenna tuner, more correctly called a transmatch, is an impedance matching device intended to make the antenna system's impedance approach that of the attached transmitter or transceiver. This assures better transfer of RF power and less of a hazard to the transmitter by high RF voltages from reflections on the feedline caused by the mismatch.

For receiving, there may be a marginal increase in signal strength and background noise when the transmatch is properly adjusted. Tuners are far broader than pre-selectors in their frequency selectivity, so they have little effect on slicing a narrow portion of spectrum out of the mire and suppressing the rest.

Q. If I were to stack two identical antennas side by side, would I add the individual gains for the total gain?
A. While this seems logical, decibels are logarithms – they indicate multiplied increases, not added increases. Assuming the two antennas are identically distanced from the signal source so that the two received signals are in phase, the second identical antenna would double your intercepted signal voltage. That would be a 3 dB increase over what a single antenna would provide.

Q. I'd like to operate two or more scanners in my car. Can I do it with just one antenna and a splitter, or do I need a separate antenna for each scanner to avoid signal loss and mutual interference between the scanners?
A. Two separate antennas for two scanners will result in the highest signal strengths and best isolation between the two scanners. It will help

prevent picking up oscillator radiation which can act like a bogus signal, locking up the scanning sequence in the other scanner(s).

If signal strengths are reasonably strong, a standard TV antenna splitter, bringing one antenna into two scanners, will work just fine. You only lose about 3 dB of signal because you divide the signal voltage in half. With more scanners there's more loss, and more chances of the interference. I'd try the splitter method first before you turn your car into a porcupine.

Q. I was given a scanner antenna and mounted it temporarily on my roof; it definitely improved reception. But due to restrictive covenants, I'm afraid to have it visible. What loss would I have moving it inside the attic at the same height?
A. If the antenna is at the same height as it would have been outdoors, and uses identical coax, the things that can degrade its performance are shielding and reflecting effects of large masses of surrounding metal, specifically heating and air conditioning ducts, house wiring, metalized Mylar insulation, aluminum siding, gutters and downspouts, and possibly the roofing tile material (depending upon its composition). I'd recommend you put it in the attic and try it; that's really the only way you'll know.

Q. I bought a desktop scanner to listen to 2 Meters (144-148 MHz). While the plug-in antenna works well for that purpose, I don't seem to hear much else, especially on 6 meters (50-54 MHz). Is a scanner antenna for 6 meters different from a ham antenna for 6 meters
A. The little plug-in antennas that come with desktop scanners are simple whips that work best in the 150-960 MHz range, but they are too short to work well on lower frequencies. Let's review some simple theory.

Virtually all antennas are fed at or near their electrical center. Even base-fed verticals use something to emulate the lower "half" of the antenna, often ground radials if earth-mounted, or metal ("ground plane") radials if elevated. In the case of the scanner, the conductive chassis components emulate a ground.

For mobile mount, the car body becomes the missing portion of the center-fed antenna. On a hand-held radio, your body is capacitively coupled to the radio to substitute (usually poorly) for the missing lower element.

Any length of metal has a specific frequency to which it is "resonant;" that is, it has a feed point impedance which matches that of the feed line, traditionally 50-70 ohms. If you make it longer, its feed point impedance rises; if made shorter, that impedance lowers.

It's an odd-harmonic, cyclical phenomenon. Once the antenna length doubles the original resonant length, the impedance is several thousand ohms, quite a mismatch to your 50-70 ohm coax! But as the length approaches three times the resonant length for that frequency, the impedance lowers again toward 50-70 ohms. That's why center-fed antennas are often used on their odd harmonics.

For example, a 50 MHz antenna will work well at 150 MHz (the third harmonic), but a 150 MHz antenna won't work as well at 1/3 its frequency range (50 MHz), and its reduced length also means less signal-voltage capture.

A thin, steel wire will receive just as well as a thick, silver pipe. Larger-diameter elements have the advantage of maintaining the feedpoint impedance for a slightly wider frequency range, that's all.

Q. What is a "Zepp" antenna? What is a composite antenna?
A. "Zepp" is an abbreviation for Zeppelin, the old German lighter-than-air craft. They used trailing antennas fed at the close end for communications; thus, a "Zepp" is a general reference to any end-fed wire antenna.

A composite antenna is simply a combination of two or more individual antennas operating together to produce either gain (focus in one or more directions) or wider bandwidth.

Q. I'd like to connect either an active antenna or large outdoor antenna to my shortwave portable, but I'm afraid of overloading the radio. Can I simply put a 5,000 ohm carbon potentiometer between the radio's antenna terminal and the incoming coax line?

A. Yes, but I'd suggest a lower resistance – say, 100-500 ohms or so. The reason for this is that the line impedance is going to be around 50 ohms, and I suspect that the high-resistance pot would be very "touchy" to tune – you wouldn't be using much of the rotation.

Q. Although my scanner antenna is advertised as being omnidirectional, doesn't the fact that it is mounted alongside a metal mast make it directional?

A. Yes; a metal mast within 1/4 wavelength of an antenna is reflective and imposes a directional effect. As to what the effect will be (enhancing, canceling, or no effect) depends upon the frequency and spacing of the element from the mast pipe, and the angle of the arrival of the signal.

But the resulting gain is only about 2-3 dB at best in one direction, and any signal cancellations (nulls) from the sides will be very sharp, so the general reception all around is still mostly non-directional.

You can always mount it in such a manner that the antenna faces the most critical, distant target on VHF-high, then listen for weak signals off the sides and back to be sure that they haven't become attenuated.

Q. I have a VHF antenna atop a 30-ft mast, but my reception is still limited. What are some general recommendations?

A. Here are a few:

    1. At least double the antenna height;

    2. Add a mast-head preamplifier at the antenna to compensate for feedline loss;

    3. Switch to lower-loss coax;

    4. Replace the antenna and/or feedline and connectors if they look corroded;

    5. Replace the antenna with one of higher gain; if you go to a directional beam, you will need a rotator to select primary targets.

Q. One of my backyard wire antennas comes into contact with a tree branch on its way to the fence where it is anchored. Should I use a stand-off insulator to keep it from touching the branch, or is it ok as-is?
A. At higher frequencies, the antenna will exhibit high impedance as you get farther from the center; if the wet tree branches or leaves touch the wire at these points, the resistive path to ground could cause minor signal fluctuation or static. It would be better to keep the bare wire from wet limbs if you notice this effect.

Q. Is there a simple way I can hook up my multiband portable to an external antenna? Can I just hook up a wire with an alligator clip to the little whip? There is a 1/8-inch antenna jack on the side of the radio.
A. An indoor antenna will pick up electrical appliance and AC house wiring noise. If you will be using an outdoor antenna with coax, definitely use the antenna jack on the side of the radio so that the coax shield is also connected as well as the center wire. If you are confined to an indoor antenna, then you might as well attach it to the whip and endure the stronger noise along with the signals.

    If your radio didn't have an antenna jack, you could attach the center conductor to the whip (compressed, not extended), and the shield to the earphone jack barrel or battery negative terminal..

Q. When I tune in my local air traffic control frequency (134.0 MHz) I hear the controller but sometimes not the planes using my attic antenna. What could cause this?
A. If they are military planes, they could be replying on a UHF frequency. Possibly your indoor antenna is suffering from reflection and/or blocking from certain directions. Finally, the responding aircraft could simply be far enough away that the signal is too weak to be heard at your location, but being heard just fine with the FAA tower.

Q. I often see antenna gain figures in either dBd or dBi; what is the difference?

A. dBd means decibels above a dipole, and dBi means decibels above an isotropic antenna—a theoretical point in space radiating uniformly in all directions.

Since a dipole focuses its power in specific directions, it has 2.15 dB gain over a spherically radiating point. That's why antenna manufacturers like to use dBi as their reference, since virtually any other antenna will have gain over it. Converting between dBi and dBd is done by simply adding or subtracting 2.15 dB.

Q. My end-fed antenna runs across my roof at a height of 30 feet. How would the antenna work if I loosely wind the wire just a few turns around a tall, two-inch diameter, PVC pipe? Would it perform better than its current installation hanging horizontally (where it acts as a directional antenna), or if it was hanging vertically (in which case, would it be omnidirectional)?
A. If hung vertically or wound around a tall pipe, it would be omnidirectional rather than have the bi-directional pattern it has now as a horizontal wire. But use very few turns around a pole so that it doesn't become a coil, resonant at a particular frequency.

Q. How long does a wire antenna have to be to be considered a Beverage? How does this affect the directivity of the antenna?
A. An end-fed Beverage antenna is longer than a full wavelength at the frequency of interest, and usually multiple wavelengths. As the received frequencies become progressively higher, their wavelengths become shorter, and more lobes develop which begin more and more to favor the ends of the wire. Aa multiple-wavelength Beverage aims at targets off its far end.

Q. I'd like to erect an outdoor antenna for my shortwave receiver, but it must not be visible. What are my alternatives?
A. There are several possibilities:
    1. Select a 25 to 50-foot roll of small-gauge wire with insulation color that matches your siding, and run it from one window to another

distant window. Connect the near end to your receiver, preferably with coax.

2. Better, select a 25 to 50-foot roll of small gauge wire with gray insulation and run it from a window or an eave to a tree limb. Connect the near end to your receiver.

3. With a rock connected to one end of an unrolled, 25 to 50-foot length of insulated wire, throw the rock as high as possible over a tree limb. Bury small-diameter coax (RG-174/U, RG-58/U) in the ground between your house and the tree; solder the center wire of the coax to the bottom of the vertical wire, but ignore the shield. Wrap the joint weather-tight with Coax-Seal™ and/or vinyl tape. 4. Buy a coax-fed active antenna and mount it with a flag hanging from it on a porch rail.

Q. Can a standard TV balun transformer be used for transmitting? How much power? Over what frequency range?
A. The larger VHF/UHF-TV balun transformers (approximately 1" diameter) supplied with TV antennas can be used for transmitting 10-20 watts intermittently, and from 2-3 MHz all the way up through 470 MHz or so, assuming you have an antenna that matches its 4:1 impedance transformation.

Q. I want to combine two remote active antennas, one that works best at VLF and the other that works best at HF so I can get the entire 10 kHz-30 MHz range. Since they both have 10 kHz-30 MHz frequency coverage, I am concerned about how to connect them to one transmission line without causing phase interaction.
A. Anytime you combine two antennas you run the risk of changes in directivity and gain. If two identical antennas are separated by enough distance and are in phase, you can get a theoretical maximum of 3 dB gain. If they are out of phase, they can null or even cancel a signal.

At the very low frequencies, due to the long wavelengths, physically close antennas – within a small fraction of the wavelength – shouldn't exhibit that problem. But at the upper HF range, separating them by only a few feet can make a difference.

To isolate the two frequency bands from one another, I would suggest putting a 2 MHz low-pass filter on the VLF antenna and a 2 MHz high-pass filter on the shortwave antenna. Close mounting shouldn't be a problem. If the supply voltage requirements are the same there should be no problem feeding them both through one common coax line. If there are equal line lengths combining the two antennas, you shouldn't experience nulling.

Q. I understand that a horizontal wire antenna receives best at a 90 degree angle to the transmitted signal, but aren't transmitted signals bounced off the ionosphere at random angles?
A. When radio signals bounce from the ionosphere, it's primarily their polarization (horizontal vs. vertical) that gets scrambled. The dominant signal strength still favors one compass direction. However, the directivity of a dipole is not razor sharp. The maximum signal 90 degrees off the sides of the wire gradually rolls off as you change the angle of the dipole away from the signal until a sharp null loses the signal off the ends of the wire.

Q. I am in the process of assembling a remote RF controlled antenna switch. It would be easiest to just use the miniature relays already on the receiver board. Do you think they would work OK for RF switching at LF, HF and VHF? I could solder coax directly to the relay and shield them with copper tape. I tested one of the relays in my coax line and did not see any noticeable loss.
A. Small relays work fine for LF right on up through the lower VHF range, but contact lengths become lossy as frequencies go higher. Without close shielding spaced alongside the contacts, the characteristic impedance changes, and the reactance and radiation can cause signal loss. Emulate the shielding as best as possible, then simply compare weak VHF/UHF signals with and without the relay, as you have done. You can't beat that strategy.

Q. What are the typical adjustable, quarter-wave whip lengths for 2 meters, 220 MHz and 440 MHz? If not extended to correct length, is there an issue to the transceiver when transmitting?

A. Mismatched lengths will compromise weak-signal reception and reduce power output during transmit. Typically, you would adjust it to a quarter wavelength for the frequency on which it is to be used:

146 MHz = 19 inches
220 MHz = 13 inches
440 MHz = 6 inches

The formula is to divide 2808 by the frequency in MHz to give you the length in inches. You can also divide 936 by the frequency in MHz to give you a 3/4 wavelength antenna with a couple of dB gain. Thus, that 19" antenna can also be used at 440 MHz. An inch or so one way or another won't make much difference.

Q. I am installing an HF antenna and a tuner for my transceiver. The ground wire is 15 feet long into a five foot ground pipe. Will this be adequate?

A. Effectiveness of your ground will depend primarily on two things: the conductivity of the soil, and the frequency you are operating on. Moist loam will work better than dry clay. The higher the frequency, the worse the 15 feet of wire will respond as a good ground.

The height and polarization (vertical or horizontal) of the antenna will play an additional part in the effectiveness of the ground. A horizontal wire should be as high as possible, at least a quarter wavelength at the lowest operating frequency.

Q. I have several wire antennas outdoors coming to a switch box via coaxial cables. Two have grounded baluns at the coax connection points using 8-foot grounding rods. Two others are grounded to a switching remote box which has its own 8 foot ground rod embedded in the soil. The coax shield is also grounded with a rod just outside my den.

Should I incorporate any gas-discharge units in the coax? Should they be installed at the remote coax cable box connections for the antennas connected to it, or in the den?

A. Yes, you definitely want gas-discharge surge protectors on the antennas. You can put them on every antenna coax line, or put one only on the coax that comes into the radio.

A trick used by broadcast stations is to coil several turns of the coax before running it into the house. This acts like a choke to retard the rapid surge. Another is to run the coax into the house through about ten feet of metal pipe which is well grounded.

But nothing will protect your radios from a direct lightning strike; that's why many seasoned hams disconnect the line from the radio during lightning storms. Some have a shorting switch that grounds the antenna, not just the coax shield.

Q. I have a switchable balun transformer for either a 1:1 or 4:1 turns ratio. It's connected between my 52 foot dipole which is fed with 450 ohm ladder line and my antenna tuner. I can't seem to tune properly in the 4:1 position, but it tunes just fine in the 1:1 position. What gives?
A. Just because you are using a 450 ohm transmission line doesn't mean that your antenna feed point is actually offering a 450 ohm impedance on your operating frequency. Additionally, even if the impedance is close, it may exhibit a very high inductive or capacitive reactance which can't be handled by your antenna tuner (transmatch) on the 4:1 position.

Q. What type(s) of antennas do modern railroad locomotives use? I haven't seen any whips on them.
A. Most seem to prefer the "blade" or "anvil" style from Sinclair Technologies due to their low profile and rugged construction.

Q. What makes a good noise antenna to be used with electrical noise-interference cancellers?
A. Much depends on the actual source of the noise. Is it from a distant power line or an indoor accessory? In many cases, a 10-20 foot length of wire strewn under a rug or run along a wall base molding is satisfactory. In other cases, you might need to run the wire out a window. It might even be possible to connect a short length of wire in

series with an isolating capacitor (.001 uF @ 600 VDC) to a wall socket. Experiment for the best noise rejection.

Q. I understand why a balun transformer can help impedance matching, for example a 4:1 balun for matching 72 ohm coax to a 300 ohm dipole, but why would anyone use a 1:1 balun?
A. A balun (balanced to unbalanced) transformer not only matches impedances, but also allows an unbalanced line like coax to be connected to a balanced antenna like a dipole. This prevents uneven currents to appear on the coax shield which would result in unwanted radiation causing distortion of the antenna pattern and/or exposure of RF power to the near-field environment.

Q. If a very long wire antenna offers no real advantage over a shorter one because both signal and noise are increased, is this true in the 200-400 kHz aviation beacon band? I've read that very long Beverage antennas are effective for beacon DXing. Don't they pick up more noise too?
A. The Beverage has large signal and noise capture because of its length, but its pattern (directivity) is sharper than that of a long random wire, so it picks up less interference and noise. Its impedance is much better matched, thus reducing line loss, and its low mounting reduces environmental noise pickup and favors direct ground wave pickup. This is true anywhere in the low-frequency ground wave spectrum.

Q. Is there a cure for the flutter I get on receiving 800 MHz signals with my mobile scanner?
A. I'm sure you get this mostly when you are driving through built-up areas due to out-of-phase, signal-cancelling reflections from buildings or even hillsides. One reader improved reception with dual-diversity antennas.

He mounted one whip on his right fender and the other on his rear deck of his vehicloe. His best results were with unequal lengths of coax to those two, non-symmetrical, antenna mounting positions. Both cables are connected to the pair side of a standard, TV-style splitter. The third port is connected to the scanner by a short jumper cable.

Q. I'd like to erect a broadband, LF/HF, receive-only antenna. I presently have a 75' random wire up about 30 feet. What is the recommended length of wire?
A. As you've probably already figured out, a random wire can have a wide range of feed-point impedances depending on wire length, frequency, and height above ground. 40-60 feet is a good average for a random wire antenna at least 25 feet high. Fortunately, moderate impedance mismatches don't matter much for receiving at LF/HF as much as they do at VHF/UHF with coax being more lossy at higher frequencies.

Q. My primary listening is VHF and UHF aircraft; is there any advantage in a discone antenna?
A. Discones were developed during WWII for ground-to-air communications, so they are an excellent choice. All should include 118-138 and 225-380 MHz.

Q. I want to improve my pirate radio catches around 6.9 MHz. I'm using a 150-ft longwire 20 feet above ground, made from common lamp cord including the lead-in. Will a "preselector/tuner" improve reception? How about cutting the antenna to a quarter-wavelength dipole?
A. Making the dipole a half wavelength (not quarter-wavelength) is theoretically an improvement since it's close to resonance and matches the receiver's design impedance. Realistically, it probably won't make much difference, but it will be shorter and more pattern-predictable. This would be about 68 feet, fed at the center insulator with any kind of fresh coax, including inexpensive RG-58/U. The shielding on the coax will reduce electrical noise pickup from household appliances.

Elevating the dipole to 30 or more feet may help by lowering the main lobe of the antenna's pattern for more distant signal detection.

Q. I would like to position my shortwave antenna for the best reception from Europe and Asia; what would the direction be?
A. Generally speaking, you'd face the broadside of the wire northeast for Europe and North for Asia. If you have a world globe, simply stretch a piece of thread or string between your location and your target to emulate your antenna. Remember, it's the sides of the antenna that finds your target, not the ends.

Q. I am planning to put a shortwave antenna on top of a 12 story building, but I would need 100 meters (330 feet) of coaxial cable. Does that present a problem? And do I need 50 ohm coax?
A. The longer the transmission line, the more loss it will exhibit due to resistive effects and absorption by the dielectric (insulation). At 30 MHz, the top of the shortwave spectrum, RG-58/U will lose at least 8 dB of the received signal, and about 5 dB at 10 MHz. RG-6/U outdoor TV coax for those same frequencies would lose less than 5 dB at 30 MHz and less than 3 dB at 10 MHz.

But if you try to use that coax for VHF/UHF applications, the difference is astronomical. At 150 MHz the RG-58/U will lose some 19 dB while the RG-6/U will only lose about 11 dB. And it gets worse as you go up in frequency from there.

Equally important is the shielding. RG-58/U may have only 70% shielding while RG-6/U is 100% double-shielded; therefore, the RG-58/U is more vulnerable to picking up electrical noise.

So far as impedance, that's of no concern for receiving over such a wide frequency range; the feed-point impedance will vary all over the place with changes in frequency.

Q. A car's battery is "grounded" to the chassis of the car. Is grounded really a good word as the chassis is in no way connected to the earth? Wouldn't "chassis ground" be a better term?
A. Yes, chassis ground is a common term used in electronics for the common return path, often the negative charge. The term "grounding" originated in the 19th century when early experiments in electrical transmission actually used the ground as part of the return path. It is still

used literally with antenna systems and commercial power distribution to prevent shock hazard and to bleed off static electricity discharge.

Q. Will four eight-foot copper rods in a 15-foot-diameter circle, fed with 6 gauge copper wire, be an adequate ground for shortwave receiving only? It is clay soil close to a septic tank drain field.
A. More than adequate. Grounding for receiving purposes is not as critical as grounding for transmitting, because we aren't looking for a low SWR for maximum power handling efficiency. A ground may not even be necessary if you don't have electrical noise interference without it. Moisture will improve the grounding characteristics. Wet, mineralized soil is best; dry sand or clay is worst.

Q. I will be installing a wire antenna for shortwave reception. How long can the ground wire be to the grounding post?
A. A ground wire on an antenna won't actually increase signal strengths above the noise floor, but under some conditions will reduce electrical interference from nearby sources. Since the coax shield is attached to the receiver's chassis, simply grounding the receiver should provide the same results.

A good ground wire should be as short as possible, and fairly large in diameter to increase the efficiency because of "skin effect" at radio frequencies. Coax shield is commonly used, as is standard-gauge house wire.

Q. How do remote antenna-mounted preamps allow both the signal and the operating voltage to be fed directly through the coax without the operating voltage damaging the receivers and scanners to which they are connected?
A. Radio frequency energy can't get through an RF choke coil, and DC can't get through a capacitor. The DC is fed from the power supply to the remote amplifier through an RF choke coil, isolating the radio frequency signal on the coax. The radio frequency signal is tapped from the coax with a capacitor, thus preventing the DC from appearing at the receiver's antenna terminal.

Q. I'm making a small indoor shortwave antenna for the first time. The radio I have has a 3.5 mm mini external antenna jack. Should the jack I use have 2 lines meaning a "stereo" plug or a jack with 1 line meaning "mono."

A. The terms mono and stereo descend from the early days of electronic music amplifiers. Monophonic meant just one speaker and/or microphone, and stereophonic meant two in order to give left and right definition to the reproduced performance as if you were in the concert hall. A mono plug has two wires attached, the barrel (ground) and the tip. A stereo plug adds the third wire, the ring. All common antenna leads (transmission lines) have two wires; with coaxial cable, that would be the shield (ground) connected to the barrel of the plug and the center conductor connected to the tip. Chances are the external antenna jack on your radio is mono rather than stereo.

Q. I need to reach a 2.4 GHz Wi-Fi tower some 15 miles away. Can I simply attach a magnetic-mount whip to my external antenna jack?

A. 15 miles is a consequential distance for reliable intercommunications with a fractional-watt device. A whip antenna is omni-directional, not directional, so the gain may not be enough. A beam antenna would offer more gain in the preferred direction.

The whip would have to be mounted on a metal ground plane (horizontal metal surface) no less than 8" in diameter; that's why it is designed to be mounted on a metal car roof.

You can predict the gain you need by seeing how far you have to drive toward the nearest cell tower to get service, realizing that half the distance is equivalent to about a 6 decibel (dB) signal improvement, plus the absorption losses from surrounding tree leaves.

Another problem is the coax cable, quite lossy at these frequencies; 1 dB per meter of length for common RG-58/U, and an antenna on the roof of the cottage would add quite a few dBs of loss.

Is cable available there? There's always satellite Internet service, although it is known to bog down with shared usage. A high-gain, 2.4 GHz beam may be your only option.

Q. I have read in the past that some folks have gotten good shortwave reception with a random wire simply lying on the ground. I tried this but my reception was noisy and signals were weak. What's the real story?
A. Similar experiments I've done in the past showed that such reception was good below about 3 MHz, but above that, it was increasingly poor due to ground absorption losses and high upward reflectivity. Dry, sandy soil would work better than moist, mineral-rich soil.

Q. I currently have my HF dipole mounted in the attic under the roof peak running power up to 100 watts. I'm planning to install a large solar array on my roof 2-4 feet above the entire length of the antenna. Wiring to the panels is encased in metal conduit. Is the RF signal likely to cause damage to the solar array or its associated electronics, and will it act like a Faraday shield on my antenna, preventing it from radiating its signal?
A. The way these cells are mounted on a metal framework, and the wiring being metal encased, I doubt that 100 watts of RF being radiated several feet away from the array is likely to induce damaging voltages.

The solar array won't act like a Faraday shield because it doesn't enclose your antenna, but there will be consequential reflective effects that will unfavorably impact your preferred horizontal radiation pattern.

Another thing to be concerned about as a ham or shortwave listener is that pulse width modulated charge-controllers, used to charge back-up batteries in photovoltaic solar panel systems, can generate a great amount of broadband RF interference during the charging process. The degree of interference will vary with the system, and the greatest amount of interference will occur during daylight hours.

You can alleviate the effects of some of the interference by placing the charge-controller as far away from your antenna and/or rig as possible, using best grounding practices and DC filtering. Make sure your installer is aware of your concerns and can plan the system accordingly.

Q. I'm having intermittent problems picking up 700 MHz digital signals on my desktop scanner with a telescoping antenna. How can I improve the reception?

A. If reception is spotty, it could be your location, the distance to the 700 MHz antenna tower, or your antenna. Since you are using a desktop scanner, it's a safe bet that you are inside your home.

Worst case scenarios are in a low (near ground) position; behind metalized Mylar insulation in the walls; nestled in among higher structures; behind a mountain or hill; farther than 10 miles from the transmitters; using a poor antenna.

First, adjust the length of your antenna to about 4 inches since that would be the resonant length of a 700 MHz quarter-wavelength antenna. You can also try 12 inches (this 3/4 wavelength sometimes works better because it has more capture area). Tilt the antenna in different directions as well.

Also try moving the scanner to a window facing the direction from which the signal is coming. A high, outdoor antenna is always best. In some cases you can use an attic crawl space.

Q. If I were to make a square of four wires for a receiving antenna, would I get signals from all four directions equally? And does it matter where I "tap" the antenna?

A. You've created a horizontal loop, and it's essentially non-directional in its uniform response from all directions. You can break the loop to feed it at any point, although it is most commonly fed at one of the corners since that's a physical support point.

The proper formula is to divide 1005 by the lowest critical frequency in megahertz to get the total perimeter in feet. This produces an impedance of about 550 feet for 1.8 MHz and higher. If you use insulated wire, reduce the calculated length by 4%.

Since this is a balanced antenna, it's best fed by ladder line (wide-spaced twin-lead). If you want to provide closer impedance matching for transmitting, you could use a common 4:1 balun transformer (or wind your own 6:1) and run low-loss coax like RG-6/U to it.

Height should be at least 40 feet for best response; the higher, the better. Lower elevation increases ground reflectivity – great for receiving overhead aircraft, but poor for long distance reception!

Q. I've installed an outdoor wire antenna, but I'm concerned about grounding. Just what do I need to do and where to I connect it?
A. In the radio lexicon, there are several meanings for "grounding." One refers to a common connection for all components that need to be eventually tied together, and that's usually the radio chassis and includes the negative power connection.

Another refers to actual earth ground, useful for two reasons: to provide a complete radio-frequency antenna system using the earth as part of the system, and to make sure there are no AC hum components to interfere with reception.

When your radio room is high above the earth, it's hard to get any antenna system improvement with a long ground wire because of the wavelength issue – the common signals won't be in phase, thus providing signal reduction. But neutralizing the hum voltage and other electrical interference from power lines and nearby accessories is still a possibility.

A good earth ground consists of at least one 8-ft metal pipe in moist earth; better, two pipes separated by several feet. The ground wire going down from your radio's chassis or coax connector (shielding) should be large, like coax braid or heavy-duty braided wire.

Even so, noise reduction isn't guaranteed; it's a hit and miss proposition, but you are always better off with the ground than without it, even if only for shock protection.

Q. Can I use satellite dish coax and connectors for scanner antenna transmission line?
A. Sure; satellite TV systems use a rugged, low loss RG-6/U and F connectors. They will work just fine.

Q. My new digital scanner is not picking up signals as well as my old analog scanner. What sort of antenna should I put up?
A. Can you hear analog signals as well on your new scanner as you did with the old scanner? In a side-by-side comparison, there should be no difference in reception since you are listening to the same signals. If there is, then I'd suspect the scanner.

Are you having to use a different adapter on your antenna cable than you were before? Make sure it's not loose and is making a good contact with the scanner receptacle.

Digital signals must be stronger than analog to be heard properly. As with TV, analog signals can tolerate some static and still be seen, but digital needs all the pulses and no noise to get a TV picture and sound.

For scanner coax, don't use RG-58/U in long runs; it's very lossy at VHF and especially UHF. Use RG-6/U outdoor TV coax or, in the more critical cases, more expensive low-loss cables like RG-8/U, Belden 9913, or LMR400. TV twinlead can also be used, but, because it is unshielded, it is vulnerable to signal-absorption losses from moisture and adjacent downspouts or metal window framing. It will also require a TV-style balun transformer with cable and connectors for the scanner.

Q. Is it reasonable to use a cellular antenna as a general-coverage, mobile scanner antenna for public safety and aircraft monitoring?
A. As a matter of fact, I frequently use cellular mobile antennas for general-purpose scanning and they are quite satisfactory. Don't expect the performance that a good, mobile scanner antenna can provide, but for general local reception, they work just fine. The more "pigtail" coils on the whip, the better the lower-frequency response.

Q. I have designed two different Yagi antennas, one with 10 dB gain and 56 ohm impedance, the other with 8 dB gain and 50 ohm impedance. Which should I go with?
A. The slight loss from an impedance mismatch of only a few ohms will be virtually impossible to detect by the receiving station, especially if you are using low-loss transmission line. A single S unit on a signal

strength meter is 6 dB, and even if you are at the fringe of marginal reception, a dB or two isn't going to make much difference. I doubt you'd see or hear any difference between the two antennas.

Q. Would shortwave reception be substantially better with an antenna 100-200 feet high than at, say, 10 feet above the ground?
A. Yes, even with 200 feet of low-loss coax. When horizontal antennas close to the ground cause reflections which make primary reception overhead rather than from the horizon; these are called near-vertical-incidence antennas.

Q. For scanner reception, what is the practical difference between RG-6/U, RG6, and RG8/U?
A. RG-6 and RG-6/U are the same thing; the "U" is sometimes left off casual references, but all coax with the military RG (radio guide) designation has the /U suffix ("utility" or "universal"). RG-6/U has less loss than RG-8/X, on the order of 3-4 dB per 100 feet at 1000 MHz. This would be discernible on weak 800 MHz signals, progressively less so as you go lower in frequency.

Q. What's inside the cylindrical "thingamajig" on some whip antennas, and what does it do?
A. It's simply a coil of wire inside a weatherproof jacket. If it's at the bottom of the whip element (base loading), its inductance neutralizes the capacitive reactance (radio-frequency resistance) that a too-short element has at a specific frequency range.
If it's between different-length segments, it's more likely a decoupling coil to isolate one section from another so each can function independently on separate bands.

Q. I currently have two VHF/UHF whip antennas on my car. I'd like to install a scanner. How can I prevent my transmitting signal from harming the scanner?
A. First, separate the two antennas as far as possible. Another protective method occasionally used is to place two parallel PIN diodes or silicon

signal diodes (1N918/1N4148 typically) in reverse polarity across the scanner antenna cable, center conductor to shield. While these may act as signal mixers in strong signal areas, producing image interference, it's worth a try.

A safer alternative would be to use a TV antenna switch or relay to cut the scanner in and out of the antenna circuit. If you're using high power, have one switch position a short circuit on the coax lead.

Q. I have a discone antenna and a good preamplifier, but I live in a rural area so that some signals are quite weak. Can I get improved scanner reception by adding a second preamp?

A. It's never a good idea to cascade two or more amplifiers. The extra gain causes more background noise and is likely to cause strong-signal overload (intermodulation or "intermod") resulting in phantom signals being heard.

Discones are known for their wide frequency coverage, but not for their signal sensitivity; they have no gain. Additionally, their directivity slants upward (above the horizon) on some bands, reducing their reception range.

If you've tested your antenna and preamp with another scanner, the only things you can do to improve reception are:

1. Get a better, high-gain antenna and point it in the direction of the most desired, weak signals;

2. Change the coax to RG-6/U (least expensive), RG-8/U foam, or LMR-400.

3. Raise the height of the antenna to be sure it clears interfering surroundings.

4. Test another scanner on the same antenna to confirm it has the same poor reception.

5. Move.

Q. I want to erect a shortwave dipole antenna in my attic. Instead of running it in a straight line, can I take each leg and run each at a different angle to each other?

A. Yes, so long as you don't close it into a V; that would make it directional, nulling out some signals depending upon frequency and bearing.

Q. What is the formula to calculate the length of a shortwave antenna wire and a mobile whip?
A. If we consider a typical, center-fed, shortwave antenna to be a half-wavelength dipole, the total length in feet is 468 divided by frequency in megahertz. Thus, a 7 MHz dipole would be 67 feet long.

Proper antenna length is important for transmitting because an impedance mismatch feeding the antenna reflects power back down the feed line, wasting the power heating the coax.

But for receiving purposes, there's no magical number since receivers are tuned over such a wide array of unrelated frequencies that no single length is best.

For VHF/UHF mobile, where the center conductor of the coax is connected to the base of the whip, the car body becomes the missing half, so the whip is only half as long.

Divide 2808 by the frequency in MHz to get the length in inches. A 19-inch, quarter-wave whip for the two-meter band is also a good length for a ¾ wavelength, 430 MHz band antenna, and for general purpose, VHF/UHF scanner listening as well.

Q. My dad used to tell me how he would alligator-clip the antenna lead from an old Philco radio to the finger hook of a dial telephone to improve reception. He said the phone lines acted as a huge longwire antenna. Is this still possible today, even without that finger hook?
A. Actually, the technique still works fine – if you can find a metal point on a modern phone to connect the antenna wire. Look for a metal screw that may go into the phone's metallic mass.

More dependably, get a phone plug to connect to the phone outlet, and using an isolating capacitor (.01 to .1 microfarad) to avoid connecting to a "hot" wire (one with voltage on it), try each of the four different wires for best reception.

Q. I presently alternate between a sloper wire antenna and a CB ground plane for shortwave listening. Why does the vertical CB antenna often "hear" better than the wire?
A. There are several possibilities, including:

    1. The vertical is omnidirectional while the wire favors the direction of the downward slope, so signals can be higher on the vertical from certain directions which are nulled on the wire.

    2. If the sloper faces your home or another source of electrical interference, that would make the noise level higher than on an elevated vertical.

    3. Signal strength readings can be misleading. What you are really listening for is signal above the background noise. A weak signal on a quiet background will be more readable than a strong signal on a strong interference background, even though the latter will produce an elevated S-meter indication because of the noise.

    4. The relative placement of the two antennas may intercept dissimilar signal strengths on different frequencies.

    5. Polarization patterns may be different for the arriving signals on each antenna.

Q. We recently installed an "invisible fence" to keep our dogs from roaming. It consists of about 1300 feet of copper wire buried some six inches in sand. If the end is connected by coax and run to my receiver, will it be an effective antenna for long, medium and short wave?
A. Yes, that 1300 foot wire would make a good receiving antenna; the lower the frequency the better. It will be outstanding for ground wave reception up to a few MHz, then gradually taper off at the high end of the shortwave spectrum.

Q. Does an elevated horizontal dipole antenna receive as well as a vertical antenna at the same location and of the same size?
A. Yes, if it is in the same polarization plane as the signal it is receiving; a vertically-polarized signal is best received by a vertical antenna if there are no accompanying signal reflections.

But realistically, shortwave signals coming from hundreds or even thousands of miles away are of mixed polarization because of the repeated reflections and distortions over those great distances.

On VHF/UHF, where you are hearing signals from much shorter distances, your receiving antenna must be mounted in the same polarization plane for strongest reception.

Q. What is the best way to connect two shortwave receivers to two antennas to minimize shortwave signal fading?
A. There are two types of diversity reception: frequency diversity, in which two different broadcasting frequencies airing the same program are tuned in on separate receivers; and antenna diversity, in which two different antennas are combined into one receiver.

For frequency diversity reception, use a TV splitter to connect one antenna to both receivers. To protect the audio circuitry, insert a 10 ohm (approximately) resistor in series with each external speaker line, then combine them into one external speaker.

For antenna diversity, erect the two antennas widely apart and as different as possible (one vertical, the other horizontal, or two different horizontal directions). Connect them both to one receiver.

Q. I am thinking about erecting an end-fed wire antenna for shortwave reception. Will reception be generally as good as with a cut-to-frequency dipole?
A. Regardless of the location of the feed point, the same-size horizontal wire will respond to signals the same, best from the sides of the wire. While there will be a theoretical loss from unmatched feedline impedance, for receiving applications, it will probably be insignificant.

Q. Would replacing my coax with 450 ohm ladder line to my 40m dipole be a big improvement? I will be using a transmatch.
A. Losses in transmission line from high SWR depend upon the frequency, amount of mismatch, length of the line, and insulation characteristics. Open-wire "ladder line" is one of the least lossy cables, if not the least, but since it has no shielding, you have to be careful not

to run it along metal surfaces, and it will pick up nearby electrical interference.

If you are getting a good impedance match and aren't hearing radiated interference from electrical and electronic home appliances or your computer, it's an excellent choice. Be sure to check the line periodically for age cracks which can hold dirt, salts, and moisture; if it shows these, replace it.

Q. Do properly installed lightning rods protect the house from lightning strikes, or does the lightning branch off to the internal wiring, plumbing and other metal within the house?
A. Lightning paths are hard to predict, but it does look for the lowest resistance path to ground. Unless you prepare a good ground field with large-gauge, conductive cabling, the lightning may not show a preference for the rods. Still, you're better off with the ground rods than without.

Q. I've got a run of 3-wire, barbed wire fence that is over a 1/4 mile in length. Would the top strand of barbed wire make an acceptable longwave receiving antenna? Or would the resistance in the barbed wire reduce the received signal beyond any benefit?
A. I would say that the length of the wire, especially if uncorroded and zinc plated, should do a good job in capturing VLF/ELF ground-wave signals. You might even consider tying the three strands together at the close end to decrease the resistance.

Q. I need a splitter for my antenna lead that passes signals down to the AM broadcast band, but most TV splitters are marked "5-900 MHz." Can they be used outside of that range?
A. Commercial TV-style splitters are marked for the frequency range that TV broadcast and cable channels need. In actual fact, most will operate down to about 300 kHz and above 1000 MHz with virtually no loss other than the expected 3 dB drop in each channel because the signal voltage is split in two.

Q. What would the effect be of taking hundreds of feet of wire and winding multiple turns around the perimeter of the attic? Would there be a benefit over having just a single loop of wire? Would the much longer length of wire enable the lower bands?
A. The aperture (capture area) of a loop antenna is defined by the dimensions of the loop, not the number of turns. Multiple turns make it a big coil, allowing it to be tuned with a variable capacitor so that it will resonate on certain frequencies. This makes it frequency selective, attenuating off-frequency signals which might cause interference from strong-signal overload.

Multiple times around the room will raise its impedance for a more efficient match to your receiver; the lower the frequency, the more turns will be required for the impedance to rise to the nominal 50 ohms impedance of most longwave/shortwave receivers. While this may deliver slightly stronger signals, it will also proportionately increase the received natural noise, so there's no real net gain of signal above noise.

Q. Why does it make a difference whether I put a preamplifier close to the antenna or right next to my receiver after the coax run?
A. Close to the antenna, you can amplify the weakest signals so they aren't absorbed by coax losses down the line. If you put the preamp at the radio end of the coax, those weakest signals will have been absorbed by coax losses, leaving nothing to amplify.

Q. If I change the 18 feet of RG-58/U cable on my mobile scanner antenna to low-loss RG-6/U, will I get better reception?
A. With such a short length, the difference would only be about 1 dB at 800 MHz, and less than that on lower frequencies, probably imperceptible even on extremely weak signals.

Q. With a good mobile antenna and quality coax, will a preamp help weak signal reception?
A. The preamp should do some good, but if the majority of your driving is in areas of strong repeaters, you are likely to increase overload interference on your scanner

Q. Should a TV-style, 75-300 ohm, impedance-matching transformer have continuity between the two windings?
A. Even though it's a transformer, some do. These are called "DC passive," and allow a mast-mounted preamp to receive its operating power through the transmission line. An ohmmeter attached between one lead of the 300 ohm side and one to the 75 ohm side will quickly reveal which type you have.

Q. I have two mag-mount, dual-band (2 m/70 cm) antennas mounted on my window air conditioner as a ground plane for a transceiver and a scanner. One antenna is tall, the other much shorter. Will their close proximity affect their pattern? If the scanner is turned on while I'm transmitting, is it likely to be damaged when I transmit?
A. If you are using a hand-held transceiver at a couple of watts and the antennas are separated by several inches, there is little danger of front-end burnout. If you are using higher power such as a mobile/base transceiver and they are quite close, I'd be concerned for the scanner being turned on.

So far as separation affecting directivity of the antennas, keep them at least a quarter-wavelength apart (19" on 2 M, 7" on 70 cm) to avoid noticeable interaction. If you want to experiment with directivity, tune in a weak signal and place the short one close to the window. Move the tall one a few inches behind the short one (away from the window) and move it around to listen for improved reception, but don't expect miracles—a couple of dBs at most.

Q. I live in a townhouse with antenna restrictions. Should I install an active shortwave antenna on my deck about 4 feet above ground level, or in my second floor window inside the house, or install an active loop indoors in my second floor window?
A. Generally speaking, an ideal HF antenna is high, outdoors, distant from the dwelling, not near power lines, and fed with coax. As each of these parameters is sacrificed, reception gets worse, either by increased electrical noise or decreased received signal strength.

For a loop to be effective, it needs to be rotated for either maximum reception or minimum noise as you change frequencies. Given the selections you have, I'd opt for the upstairs active whip.

Q. When using an antenna tuner (transmatch) with a transceiver to make the entire system matched at 50 ohms, how do you know how much of your RF power is actually getting to the antenna and not being wasted heating up the coax? Where do I place an RF wattmeter to determine my radiated power?

A. The transmatch becomes part of the antenna system, providing "conjugate matching" between the transmitter and feedline. This does nothing to change the mismatch at the antenna feed point; it only means that all the reactances of the antenna, feedline, and tuner now add up to 50 ohms resistance as "seen" by the transceiver.

The lossy heating of the coax occurs at high voltage intervals every quarter-wavelength, and is produced by intersecting waves of forward and reflected RF power forcing current through the insulation. All of the RF power, reflective or forward, that makes it to the antenna is actually radiated.

If you place an RF ammeter between the output of the transmitter and the input of the tuner, it will display all the power, forward and reflective, that is going into the antenna system, including that which is wasted heating the resistive elements (coils, coax, traps, antenna element, etc.).

If you place it at the antenna junction, it will display what's being radiated, including the infinitesimal amount wasted by the resistive antenna element. The difference in readings between the two points will be mostly the line loss.

Q. Will rain increase the range of low frequency radio signals that rely on ground wave such as the AM broadcast band and lower? I would think that rain would increase ground conductivity.

A. Although the water can increase surface conductivity, the raindrops/mist can attenuate signals, and if it is accompanied by lightning, that adds to the background noise. Unless it's a saturating rain, the surface improvement would be rather superficial, because the water itself is actually a very poor conductor; it needs dissolved salts to ionize, increasing the overall conductivity of the first few feet of ground. So just because it has rained, don't expect much improvement in LF reception!

Q. I notice that tuning at lower power levels and then increasing the power sometimes causes the SWR (standing wave ratio) to be higher. Is this because of the higher power (and thus higher voltage) forcing higher current through the insulation (dielectric)?
A. The radiated energy is probably making its way back into the transmitter, causing an inaccurate reading. If the problem disappears when a shielded dummy load is used, then we know that the above is correct because there are is no reflected energy to produce standing waves.

With high SWR, the reflected RF interacts with the voltage vs. current properties of the meter's detector diodes and the coupling coefficient of the directional coupler, thus giving erroneous readings.

Q. If the only way you can know for sure how much power actually reaches your antenna to be radiated is to place a wattmeter at the antenna feedpoint, does that mean that there is nothing I can put down at my rig to give me an accurate reading for the final radiated power?
A. That's exactly what it means, and why so many hams and professional installers use commercially made antennas with known matching characteristics.

Q. If I am using a 20 meter antenna for 15 meter transmitting, is most of my power lost in heating the coax feed line, or is most of the power present at the feed point and being radiated?
A. If your antenna is a conventional dipole or single-band beam, this will represent a bad impedance match with a great deal of reflected power on the feedline as indicated by a high VSWR. In that case, most of the power is wasted heating the transmission line if you're using coax, Many old-timers and experienced, multi-band, field-day operators

use open-wire feeders – no dielectric losses and eventually all the power gets radiated.

If the VSWR gets too high, modern transceivers shut down their power to keep the high voltages from burning out a transceiver's final amplifier components. But keep in mind that even if the radiated power is knocked down to only 25%, that's still only one S-unit (6 dB) loss.

Q. I would like to replace the short whip that comes with my desktop scanner with a better one. What do you suggest?
A. Replace that short rubber ducky with a longer, frequency-resonant one, or a length-adjustable, telescoping whip.

Q. What effect will using an 800 MHz antenna have on listening to VHF signals?
A. Not even considering the impedance mismatch losses, the short length of this 800 MHz antenna will exhibit reduced performance at VHF because of the smaller aperture (signal-gathering size).

Q. I'm thinking of mounting a beam antenna and a vertical antenna in parallel for improved reception. Is this a good or a bad idea?
A. When a signal arrives obliquely to such an array, one antenna is farther from the signal than the other, and the signal strengths may either add or subtract, depending on the wavelength and the angle of arrival due to phase differences when they combine.

If you can rotate it, that may work, canceling signals in some directions and at some frequencies—all unpredictably, especially since they are different in design.

Even so, in the directions that they are adding their signals properly (in phase), the maximum gain would be only a few dB, but when they cancel out of phase, they completely null out the signal.

Arrays are always done using identical antennas for predictability, and usually over a narrow band of frequencies. They are very directional with deep nulls off the sides.

Q. If I erect an inverted L antenna running 30 feet up, then 45 feet horizontally, will it pick up signals from all directions?
A. Yes, but because it's not entirely vertical, there will be some frequencies where there will be some signal cancellation from certain directions. The horizontal section will be directional off the sides, and phase differences between the upper and lower sections will produce some signal nulls from certain directions.

But the non-uniformity will be most apparent on the higher frequencies because of the shorter wavelengths for that antenna length. When you listen to the lower frequencies, the antenna will be very suitable for reception in all directions for the majority of your listening.

Q. I live 5 miles from Huntington, WV and listen to fire and EMS frequencies in the 155 and 450 MHz ranges. My antenna is on a mast two stories above ground level and picks up 155 MHz signals 25-30 miles away, but 460 MHz reception is spotty only 7-8 miles away. I've tried moving the antenna and combining multiple antennas, but nothing seems to work. Do I need a booster?
A. Since I don't know what kind of scanner, antenna, or coax cable, or its length that you are using, it's difficult for me to suggest improvements. But here are some suggestions:

Try another scanner on that antenna to verify that it's not the radio.

Check out the coax and connectors to be sure they are in good condition (not old, weathered, or moisture-intruded), and are making good connections with your scanner and antenna.

Check the center hole of the BNC connector to be sure the center blades are not splayed out. Is the UHF station you want to hear using a repeater, or are they just a low-power base station for their immediate city limits?

The coax should be RG-6/U since it has much lower loss than RG-58/U, especially at UHF.

If your antenna is a discone which has no gain; try a gain antenna or a directional beam. If these measures don't provide better reception, you might try a preamplifier ("booster").

Q. I've heard that you can protect the input circuitry of a receiver from overload damage with a pair of diodes connected in parallel, but cross polarized, across the antenna jack. But wouldn't this cause spurious signals from mixing strong signals?

A. This is apparently a myth. Many manufacturers offer PIN diodes for just such an application. To test this, I connected a pair of cross-polarized diodes across the input of a spectrum analyzer connected to a receiving antenna and transmitted 100 watts into an antenna just a few feet away.

No mixer products were seen anywhere in the spectrum. I repeated the experiment using a variety of rectifier, germanium and silicon diodes. The choice of diodes should be those with the lowest junction capacitance to avoid signal attenuation at VHF/UHF frequencies. PIN diodes are well suited for this use.

Q. What type of antenna is used for submerged U.S. submarines to receive very low frequency (VLF) signals, and where are these transmissions sent from?

A. When a submarine is surfaced, it uses conventional antennas including short, tuned verticals, patches, and dishes for two-way communications. Frequencies throughout the spectrum from VLF to microwave are employed for ship to shore, ship to ship, ship to air, and satellite comms.

When submerged, it's a different story. Seawater can only be penetrated a few feet by normal radio frequencies – the lower the frequency, the deeper the signal can be received from ground stations. Command instructions are received by a buoyed antenna which may remain just below the surface to avoid detection. Trailing cable antennas are also employed.

The primary U.S. Navy VLF station is NAA located at Cutler, ME and running nearly two million watts of power at 24 kHz. The former, extremely-low-frequency Project Sanguine was replaced by Seafarer (Project ELF) operating on 76 Hz. Transmitting stations were built at Clam Lake, WI and Republic, MI. Submarines did not transmit back. That system, with its 32 mile antenna, was dismantled in 2004.

Q. Does the quarter-wavelength minimum spacing between antennas apply to both passive and active antennas?
A. No, either will experience a "beaming" affect on the pattern, both for receiving as well as transmitting. The quarter-wavelength effect is more noticeable the higher in frequency you go, because the separation becomes closer and the masses of interactive metal occupy a greater percentage of that smaller environment.
    Their relative physical lengths are also important. If they are similar in size (within a few percent), or a whole-number multiple of that length, there will be more pattern skewing.

Q. Should a random-length, shortwave receiving antenna be end fed or center fed?
A. It really doesn't make a lot of difference, but it's a question of impedance matching. At shortwave frequencies, a short wire will have a closer impedance match to 50-70 ohm coax at the end of the wire. Longer wires have higher impedance on the ends. For receiving purposes, you'd probably never hear the difference either way you attach it.

Q. Weak-signal operators use horizontal beams to take advantage of something called "ground gain." What is this?
A. By tilting a horizontally-polarized, tower-mounted beam antenna slightly downward at the correct angle(s), a ground-reflected signal may combine in phase with the direct wave (space wave) to create an increase in signal strength, sometimes as high as 5-6 dB under ideal, flat-ground conditions.

Q. I erected a 60 foot tower with a high gain HDTV antenna and pre-amp located at the top of the tower. While my FM reception is really good, a local station on 102.7 MHz overloads my tuner so that I hear them on several spots across the dial. Is there a sharp, one frequency notch filter that would only attenuate the offending signal without impacting other FM frequencies?

A. Chances are that it's the antenna preamp, not the tuner that's getting overloaded. Take that out of the system to see if it takes care of the problem. If it does, then see if you can live with the signal levels of the other stations since no notch filter is going to notch out just 102.7 without spreading its attenuation several megahertz up and down the dial.

Move the beam around until it finds a null for the offender, reducing the images. Just hope that it's a direction that still allows reception of desired signals.

As last-ditch effort, you might want to try a receiver with better dynamic range.

Q. I would like to connect two receivers to the same antenna. Should I use a two-position antenna switch or a two-way splitter?

A. The main advantages of the switch would be much greater isolation between the receivers, thus minimizing any oscillator radiation from one getting into the other; and complete, no-loss signal strength from the antenna to the receiver of your choice, assuming the switch is lossless at the higher UHF frequencies.

A second advantage to the switch is that a splitter will lose 2-3 dB of your signal strength since the original signal power is now divided in half to feed both receivers.

Of course if you intend to use both receivers simultaneously, the splitter is the obvious choice, and the slight loss will be noticed only on fringe-strength signals.

Q. Why do fractal antenna designs only seem to be for professional uses and not for hams, SWLs and scanning enthusiasts?

A. The primary advantage of a fractal antenna is its ability to be reduced to as much only 1/4 the size of a conventional dipole or loop antenna at any given frequency without a substantial reduction in gain, and while still maintaining reasonably wide bandwidth.

They are more difficult to design and manufacture than a conventional antenna, so unless there is a critical application requiring the fractal approach, there's no reason to replace the conventional, lower-cost antenna.

The popular log-periodic dipole array is actually a fractal design, but why replace it with a more elaborate configuration for amateur radio applications?

Fractal geometry is difficult to design and construct, so its commercial applications are primarily in the UHF/microwave region where it can be etched on a substrate.

Q. Does a long-wire antenna receive signals from the sides or from its ends?
A. The term "long wire" is frequently misused; it actually means a wire longer than a whole wavelength at its operating frequency. Thus, a half-wavelength wire on 10 MHz is actually 1-1/2 wavelength long-wire at 30 MHz.

Fractional-wavelength antennas receive and transmit signals off the sides (at right angles to the wire axis), but the higher the frequency, the more that pattern drifts toward the ends.

Q. In my apartment I'm limited to an indoor antenna, but I can run an antenna wire along a 120-foot roof truss. Should I also wrap it back around several times like a giant loop? Do I need a tuner?
A. If the wooden roof-truss system doesn't have large metal masses like heating/air conditioning ducts or electric wires running parallel within a few feet of it, it should work just fine.

If your main interest is shortwave, then you really don't need more than about 50 feet of wire. If you use considerably more than that, signals may be slightly stronger, but so will the background noise, so you won't really gain anything.

Don't worry about the gauge of the wire, or whether it's stranded or solid; anything will work well. But do use coax cable down to your receiver, otherwise you are likely to hear a lot of electrical interference from household appliances and accessories.

A tuner (transmatch) is unnecessary. In cases of strong signal interference, a preselector should help.

Q. A local AM radio station on 1350 kHz has a strong third harmonic on 4050 kHz, but not on the second harmonic, 2700 kHz. Shouldn't the second harmonic be stronger than the third?
A. Although harmonic signal strengths diminish the higher they are in frequency, odd harmonics are normally stronger than even harmonics because the impedance match of transmitting and receiving antennas is closer on odd harmonics, so there is less line loss.

Q. Does the law of diminishing returns apply to antenna height? Are there conditions for which increased height won't increase reception?
A. The answer is a qualified "yes," depending on the frequency and the coax. At lower frequencies like HF (under 30 MHz – the lower the more so), signals have a tendency to follow the curvature of the earth, so the height isn't all that important. At higher frequencies (VHF/UHF), signals are more line of sight, so the higher the better.

A rule of thumb is that when you double the height, you gain 3 dB (half an S unit). To get the whole S unit (6 dB), you'd have to quadruple your height. If you started at ten feet, you'd have to put up a 40-foot tower for that S unit; if you started at 40 feet, you'd have to go 160 feet.

But the higher you go, the more coax you need. If you use RG-58/U at those higher frequencies, going higher actually attenuates your signal. That's why cell phone towers use hard line, and why you shouldn't use RG-174/U for VHF/UHF even in a short mobile installation.

Q. What are standing waves?
A. If there is reactance (capacitance or inductance) at an antenna's frequency, it needs to be canceled by a transmatch ("tuner"). Such reactance builds up large voltages on the transmission line; we call this mismatch "voltage standing wave ratio," or VSWR, often shortened simply to SWR.

These higher voltages cause power loss by heating the insulation in the feedline, and can also reflect back into the transmitter, damaging components in the final amplifier stage.

Q. If I walk around the room with my portable radio, sometimes the signal will improve, and sometimes it gets weaker; I may even cause interference on the signal if I hold the whip with my hand. What causes these effects?
A. Different parts of your home have different amounts of metal in them (wiring, appliances, metalized insulation, reinforcement rods and screens in the walls, metal siding, heat/air duct, etc.). These can block incoming radio signals, and also scatter signals which recombine out of phase, minimizing the signal in some locations.

Holding the whip antenna adds your conductive body to the length of the antenna, picking up more signal strength, including nearby interference (fluorescent lights, appliances, etc.)

Q. Can I run an antenna wire across a wooden ceiling beam, then turn it 90 degrees toward the wall, then run it back, making a "U" shape?
A. The best configuration is straight, second best is L-shaped, and worst is the U shape, because when it folds back on itself, signals from some directions will actually be reduced because the same signal is received on the two wires running in opposite directions, so the two signal voltages meet going in opposite directions (out of phase) and subtract rather than add together.

An alternative is to run the wire in three dimensions – along the wood horizontally, then the L, still horizontally, then up or down for the remainder.

Keep in mind that indoor shortwave antennas are vulnerable to picking up all sorts of environmental electrical noise interference, and signals are also attenuated by metalized insulation, wiring, heating/air ducting, metal siding, etc.

Q. Is there any difference in mounting a VHF marine whip upward or downward on my boat?

A. Not if it is mounted in the same space and the coax doesn't dangle within 18" of it. But keep in mind some other considerations.

Unless the whip is mounted on a metal surface or has ground plane elements at its base, you have only half an antenna; a metal mounting surface (or base elements on the antenna) comprises an active part of the beam-forming system. Lack of this metal counterpoise will reduce the range of communications.

If you are mounting it under a metal roof, are the cabin walls also metal? If so, even with window ports, this partially emulates a "Faraday shield," a metal enclosure which blocks radio waves. Find a location which allows the whip to be mounted either up or down on a substantial (several square feet in area), horizontal metal surface.

While a sheet-metal plane is ideal, the middle of several feet of metal piping or tubing will work well, provided the metal base of the whip mount makes good electrical connection to the metal support.

In the absence of such a location, you can mount a ground plane antenna atop a metallic or non-metallic mast, or even alongside a non-metallic mast, or at least 2 feet away from a metallic mast for VHF marine frequencies.

Q. What would be a good way to ground my portable shortwave radio to see if it cuts down the electrical noise I'm receiving with my indoor wire antenna?

A. It's questionable whether earth-grounding the chassis of your radio will reduce electrical noise interference, but it's worth a try. The easiest way to do it would be to connect the ground wire to any metal jack on the radio, like the earphone plug or the outside shell of the external antenna jack.

If you are using a plug into the radio for your wire antenna, you could use the barrel of the plug for your ground connection since the tip of the plug is the "hot" antenna connection.

Try grounding to a metal water pipe if you have one – most are plastic now, or to the chassis of another electronic accessory, or a short wire to an earth ground (a metal pipe or rod at least four feet into moist soil).

Q. When I install an antenna, should I use dielectric grease or an anti-oxidant on the connector terminals, or would this affect the performance of the antenna

A. Anti-oxidants specifically made to prevent weather oxidation of electrical terminals will not interfere with reception so long as the application merely coats the outside of connections and doesn't prevent secure electrical connections between the antenna and the lead-in.

Q. What should I use for the noise antenna on my noise canceller?

A. The key here is to separate the desired-signal antenna from the noise antenna so that you don't cancel both the signal and the noise. Put the noise antenna near the noise source; with that canceled, the only thing left is the desired signal from the receive antenna.

Mount the serious antenna as high and away from power lines and household noise as possible. You could carry a portable shortwave radio with you to listen for the minimal noise spot to install the antenna.

For the noise antenna, run a random wire around the edge of your ceiling or floor for 10-20 feet to pick up enough ambient noise to cancel the lesser noise heard on the receive antenna.

Q. How often should I replace my outdoor coax cable? When I do, should I replace it with LMR-400?

A. The general rule of thumb is to replace outdoor coax every five years. LMR-400 is low-loss cable, but only by 2-3 dB at 900 MHz, and it's considerably more expensive. You would only see very minor improvement on the weakest signals at that frequency, and less at lower frequencies.

Q. I remember a gadget that you plugged into an electrical outlet to use the household wiring as a substitute for an outdoor TV antenna. Did these things actually work?

A. They didn't work worth a hoot or they'd still be on the market. Some of the deficiencies include:

1. Household wiring isn't resonant on any particular TV frequency for proper impedance matching;

2. Wiring is electrically long at TV wavelengths, rejecting signals broadside to the wires;

3. Incoming signals were blocked by household aluminum siding and metalized Mylar insulation;

4. Since they were connected to appliances all over the house, they were rife with electrical interference;

5. Household wiring is routed near metal ducting and metalized insulation, shield it from signals.

Q. The Slinky® toy is periodically resurrected as a portable dipole for ham and SWL use. Can it be assumed that its performance is really the same as a conventional wire dipole of the same length?
A. The Slinky® is a coil; its inductance varies with frequency depending on how compressed or stretched it is. If we aren't concerned about transmitting, then yes, its aperture (signal-gathering length) would be roughly equal to a straight wire antenna of the same length.

Q. I am using RG-174/U coax to an antenna. What do these letters mean, and can I use the coax outdoors?
A. RG stands for "Radio Guide," a reference to coaxial cables developed during WWII. U means "Universal," referring to its general applicability. RG-174/U is outdoor-rated, however, like any coax, sunlight will gradually degrade the vinyl insulation. RG-174/U is quite lossy the higher you go in frequency, so use it only in short lengths and preferably under 30 MHz.

Q. I am trying to receive an 800 MHz trunking system that's about 90 miles away, but can't hear a thing on my beam antenna. What might be the problem?
A. Signals from 800 MHz base stations reach about 50-75 miles maximum under the best conditions. LMR-400 coaxial cable is the best choice among common cables, but lower-cost RG-6/U coax is usually adequate for runs under 100 feet.

Other degrading conditions include:
- Intervening buildings, hills, trees
- Losses in a long length of cheap coax cable
- Defective balun transformer
- Wet weather
- Trying to hear digital communications on an analog scanner
- Desensitization of the scanner from nearby, strong-signal overload

Q. I have a length of RG-6/U coax with F connectors; I have a PL-259 adaptor on one end and a BNC adaptor on the other. When I use it between my receiver and the antenna, I lose all signals. How do I determine if there's a short circuit in the line?

A. First, remove the adapters and make sure the center wire in the cable extends at least to the outer edge of its F connectors. If it doesn't, use a pair of needle-nose pliers to pull it out further. Use an ohmmeter set on low ohms (or a battery and bulb or LED) and proceed with the following tests:

Test 1: Connect the test prods between the outer shell and the center wire of the connector. You shouldn't get any movement of the ohmmeter; if it shows a measurement, there is a short circuit in the cable, most likely where it attaches to one of the F connectors. Note the reading in ohms, and repeat the measurement at the other F connector. If the reading is higher, the short is at the first connector; if lower, the connector you are currently measuring.

Test 2: If there was no reading during the first test, then short circuit one F connector with clips on a test lead, and take a reading at the other F connector; it should show just a few ohms of wire resistance. If there is no reading, this would indicate a break in continuity in the cable, possibly at one of the F connectors.

Test 3: If previous tests are not definitive, remount each adapter, one at a time, and repeat the tests.

Q. What is a Faraday shield?

A. Since radio waves cannot penetrate metal, a receiver or transmitter with its antenna located in a metal enclosure cannot communicate with the outside. Such an enclosure is a Faraday shield, named in honor of a pioneer in electrical research, Michael Faraday.

Equipment designers, manufacturers, and laboratories will often utilize metal-enclosed rooms to avoid interference with tests and measurements. That is known as a Faraday cage.

Section 4: INTERFERENCE

Q. What might be the cause of the strong buzzing noise interference I hear every 64 kHz (64, 128, 192, 256, 320, 384, 448 and 512 kHz) in VLF range of my receiver? I have my computer system shut off.
A. Most likely a switching power supply plugged into the wall for a cell phone, computer, or some other small accessory. The heavier transformer wall warts don't do that.
   One way to find it is by carrying a small AM radio tuned to one of the spurious signals around the house listening for the signal to become stronger.
   You can also switch off and on each of the circuit breakers on your AC power panel listening for it to stop to isolate which circuit has the offending device.

Q. I occasionally hear AC hum on some shortwave stations. I tried using battery power but that didn't help. What causes it and can I do anything about it?
A. Are you using any other accessory connected to your radio such as a preamplifier? If AC operated, they are a common source of common-mode hum. Battery power of the accessory helps. Another dreadful source of such noise is the modern switching power supply (see previous question and answer).
   There is a possibility that you are receiving actual AC line noise impacting your antenna right along with signals. Depending on where the line noise originates, it sometimes arrives in harmonics of the lower frequencies so it will effect some signals and not others. Listen for the noise on spots between signals. If you hear it there, that could be the problem. Try briefly grounding the radio to see if the interference changes.
   Finally, there's always the possibility that the transmitter is emitting the hum. This is true with some of the old equipment used in Cuba and elsewhere.

Q. I have two portable radios connected to outdoor wire antennas, each wire in a different direction. At night I hear a prominent hum interfering in my AM broadcast band reception. Is there a device that can remove that?

A. If the hum is actually a raspy "buzz" sound, I would suspect that you are getting power-line noise from nearby. Does it go away if you've had a rain? That would pretty well cinch that answer. If that's the case, you would need to drive around using your AM car radio to see where the interference peaks and report it to your utility company.

There is also a condition known as an "AC hum loop" which may result from improper grounding. Are you using a ground? Try connecting and disconnecting the ground to see if it changes the noise level.

Check to see if you still hear it with the AC adaptor unplugged and the radio operating on its batteries; that would indicate whether the noise is coming in through the AC lines, or may even be generated by the power supply, especially the notorious, switching type. Do you hear it on both radios? Is there a difference in noise level if you switch antennas? That, too, would suggest that the noise is coming in through your antenna(s).

Do you hear it without the external wire antenna connected? If so, does it change level as you rotate the radio on its base, changing its directivity? This would also indicate reception of a nearby noise source, including a household appliance. As mentioned in a previous Q&A, you can carry the radio around using battery operation to locate such a household-appliance noise source like fluorescent lighting and switching-type power supplies.

Q. I have a portable shortwave receiver to which I just added an 80' long wire antenna coupled with another long wire. Unfortunately, a 1,000 watt AM tower is located about 1300 feet away, and the radio is severely overloaded throughout the shortwave broadcast bands. Would a high-pass filter likely do the job?

A. I would suggest two things, the first of which is to shorten the long wire to about half its current length. With today's radio sensitivity levels, wires no more than 30 feet work just fine for shortwave reception.

The second would be to direct the axis of the wire directly toward the broadcast antenna, minimizing its signal pickup from that direction. Still, such a behemoth signal may well need further attenuation, and a high-pass filter should do it.

Q. Does an active antenna reduce HF interference? I have quite a bit of interference with my home random wire.
A. I'm afraid that electrical noise is a part of the electromagnetic spectrum, and if it's present, a preamplified antenna is going to boost it just like any legitimate signal.

Some folks resort to loop antennas in an attempt to null the source of interference, others elect to install a noise canceller to blank out the noise.

Always use coax feed line to an outdoor shortwave antenna mounted as far as practical from the dwelling and power lines.

Q. I'm experiencing severe electrical interference from my house wiring on my shortwave receiver. Can I simply use a capacitor across the AC line to reduce it?
A. First, try to determine what is causing the buzzing in the AC line; it could be a loose connection sparking in the wiring, and that's more than an interference hazard!

If you don't know which outlet is most affected, use a battery-operated AM radio as a probe. Set it near the wiring to pick up the noise and switch the breakers in your panel on and off one at a time to isolate the circuit.

After you know which circuit is affected, unplug anything in those outlets until the sound goes out. You have found the culprit – either the accessory or the wall socket.

If you have found the offending accessory (motor, dimmer, touch lamp, etc.), yes, you can usually reduce the interference by connecting a 0.047 (.05) uF @ 600 V capacitor across the AC connection in the device. The connection should be from the hot line to the ground. In a three-wire cable, it's often better to use two capacitors, one from the hot lead, one from the neutral, and both to the ground wire.

Q. I think I have found the cause of a "buzz saw" interference on some shortwave frequencies. I unplugged all indoor sources—computer, printer, TV, and even my cable box—and carried a portable radio around the house as a signal tracer. Nothing made a difference.

I then held the radio next to the outside junction box for our cable TV service; it's located close to all my co-ax lead-ins. The sound was loud and clear. Does this make sense?

A. It sure does. Listen to the buzz and see if its characteristics change every few seconds; if they do, you're hearing the video scan of the cable signal; as the picture changes, so will the sound of the scan.

Q. We use an AM radio in a steel-roof building, but interference from the ceiling fans with an adjustable rheostat is very bad. In addition, the thermostat in my fish aquarium makes horrendous noise on my shortwave radio. How can I fix it?

A. If you are using commercial wall speed control/dimmers, you can often reduce some of the interference by installing a capacitor, typically 0.047 (0.05) microfarads at 600 working volts (600 WV) across the control terminals that feed the fan or lights.

Be sure to shut off the circuit breakers to these controls before attempting the modification, and dress the leads with electrical tape to prevent a shock hazard. For the aquarium, attach the capacitor as close to the thermostat contacts as practical.

Q. Our church service uses a wireless mike on 72.9 MHz and there is some sort of strong interference that interrupts the sermon. Any suggestions on how to find the offending device?

A. Acquire a scanner that includes 72.9 MHz reception and listen for the signal to get stronger at various locations. If the signal only occurs when the congregation arrives, it may be one of them. Set up the scanner at the entrance and watch for the signal to appear when the guilty party walks in.

Q. I'm trying to hear a distant FM broadcast station on 92.7 MHz, but a local station on 92.5 MHz blocks it. Is there an antenna filter available that would block the offending signal, and can I simply wrap a few turn of wire around my whip to couple an outdoor antenna to it, since there's no external antenna jack?
A. Because the offending signal is so close in frequency to the desired signal, an external filter won't work; they aren't sharp enough. Such an attempt will merely decrease the signal level on both frequencies.

If the whip is capable of being swiveled left and right, you might try different angles. Also try moving the radio to different locations in your home to see if various places will shield the radio from some of the interfering signal strength.

Place the radio alongside a large metal surface like a filing cabinet, refrigerator, a washer or a dryer. Move it around while listening it to see if you can get some reflective cancellation of the undesired signal.

Q. When using a linear amplifier on a transmitter or transceiver, where do you put the low-pass filter to reduce spurious signals, before or after the linear?
A. Reasonable-cost components are more readily obtainable for low power than high power, so put it between the rig and the linear amp. The filter will to suppress unwanted, spurious signals by a fixed number of decibels, so it really wouldn't matter where you put it in relation to the amplifier.

Q. The power line near my home is causing electrical interference on my shortwave receiver. The power company drove by with a truck, listening for it with a whip on their truck, but didn't come up with anything. Is there a better way to find the source of the noise?
A. Drive down the road with your car radio tuned to an empty AM frequency and listen for the noise. You can also find these offending

poles with a hand-carried transistor AM radio. Hold the radio so you're looking at its side, not its front, and turn around until it the noise is sharply reduced; that's the direction of the noise source.

Kick the pole several times to see if the noise changes. Note the number on the pole and the address of its location and notify the power company.

Q. Automotive battery charger instructions say to connect the red (positive) wire to the battery + terminal, then connect the black (negative) wire to the chassis, not the battery's negative terminal, to prevent any spark which might ignite the hydrogen gas near the battery vents. Wouldn't it be just as safe to simply not plug in or turn on the charger before connecting both wires to the battery terminals?
A. Yes, but what if you had merely thought it wasn't turned on? By avoiding that second connection to the battery terminal, there's no way a spark could occur next to a vent.

Q. My outdoor antenna is connected to my portable multiband radio by TV-style coax which should be well shielded, but when I run the coax near a fluorescent light fixture, even with the outdoor antenna disconnected, the electrical noise from the light is picked up. How can this be?
A. A properly shielded system requires that not only the center conductor of the cable is shielded by the braid or foil, but that the receiver components are shielded as well.

When the far end of the cable is unconnected, and the receiver itself is in a non-shielded plastic box, the entire length from the far end of the cable, through the radio, and on down the power cord, becomes one giant antenna, vulnerable to intrusion by conducted electrical noise.

Another very real possibility is line-conducted noise coming in through the power cord. You can verify this by unplugging the radio from AC and running it with batteries.

And one final thought: Don't run the coax near the fluorescent light!

Q. I'm new to shortwave listening, and I'm hearing a lot of electrical noise on my portable. The noise disappears when I shut off the circuit breaker to the house, but that's not very convenient. Would putting up and outdoor antenna connected to the radio by coaxial cable solve the problem?

A. It sure wouldn't hurt. If you unplug the radio from the power line and operate it from batteries, do you still hear the noise? If so, it's being picked up by the antenna rather than through the AC line.

An outdoor antenna, fed to the radio by shielded coaxial cable, will definitely improve the situation. It will make signals stronger and will shield the incoming signals from the indoor noise sources.

Common sources of indoor electrical interference include fluorescent lights (switch them off to see if that's the problem), and control circuits in power supplies and appliances.

Section 5: MISCELLANEOUS

Q. Can hams and other electronics hobbyists transmit their walkie-talkies at air shows? Airports? Hot-air balloon events?
A. Yes. The general rule here is that such intercommunication is allowable unless it is expressly prohibited, such as construction sites where accidental explosive detonation could be triggered, or government and scientific listening posts where all environmental electronics are carefully controlled. Such protected locations will be posted.

Q. I am interested in listening to AIR New Delhi (Vividh Bharati) with its 500 kW power on 9865 KHz from New Delhi, India. Will I able to hear it with my shortwave radio?
A. The answer is the same for all shortwave listening:
    1. The season
    2. The time of day
    3. The directionality of their antennas
    4. Co-channel interference
    5. Local electrical interference
    6. Your receiver's sensitivity and selectivity
    7. Your antenna's location and directivity
    8. Their broadcast schedule

Q. How did early radio networks share programming when telephone lines had poor quality?
A. During the 1920s, networking was, indeed, done over telephone lines. If this was impractical, secondary stations could rebroadcast received signals like a repeater, or play delayed programs on transcription records (16", celluloid-coated, aluminum discs played at 78 RPM).

Q. A ham friend of mine who lives in a mobile home community says that he could never run a 1 kW HF amplifier at his place because he only has 15 amps maximum total AC house current available from his

120V outlets. Couldn't he simply use a 13.8VDC @70 amps transistor-type amplifier with an appropriately rated power supply instead?
A. A 70 amp, 13.8 VDC power supply would deliver nearly 1000 watts, but efficiency is only about 75%, therefore it would take over 1300 watts to operate it.

It also needs an AC/DC power supply to convert the 120 VAC house wiring to 13.8 VDC, also at 75% efficiency, so the total efficiency of both power supplies together is only 56%.

Using the two power supplies would take almost 1800 watts of AC power to deliver the same power that the original 120 VAC would have provided directly.

Q. I have a choice between two high-voltage transformers for a climbing-spark Jacob's ladder; one is 10 kV, the other 12 kV. Should I choose the higher current or the highest voltage for the best spark gap?
A. Use them both. Connect the primaries in parallel and the secondaries in series and you'll have 22 kV! But the primary leads have to be properly phased so that the output voltages add to each other rather than subtract. You won't know without actually drawing an arc and seeing if it's longer or shorter than with just one transformer, and that's OK to do—briefly.

During my careless youth I acquired the transformer from an old Xray machine—100,000 volts worth! It was immersed in oil and covered with a layer of beeswax. The first time I tried it, it threw a 4" arc and climbed up the ladder until it was about a foot across. Unfortunately, that short adventure spelled doom for the transformer—all I had left was a smoldering can of oil and beeswax.

Q. I recently purchased a digital camera and a laptop which came with small, lightweight battery chargers. Since they don't have a heavy transformer, how do they work?
A. These **switching-type** power supplies use solid-state electronics to reduce and rectify the higher AC line voltage down to lower voltage to operate the equipment and charge the batteries. Their only drawback is that if not properly designed (the cheaper ones), their switching

circuitry generates severe hash noise interference in nearby radio receiving equipment.

Q. I know that radio waves in free space travel at the speed of light, but how fast do electrons flow in a wire?
A. It's not the electrons, but the wave of electromagnetic energy that travels along the wire at nearly the speed of light, just as with radio frequency energy in antenna elements in radio frequency (RF) systems.

In direct current (DC) the electrons do move through the wire, but at a snail's pace—about 2 feet per hour. In household alternating current, they never leave home, vibrating back and forth 60 times per second.

Q. Electrical current is measured as the number of electrons passing a fixed point in one second. Where do these electrons come from?
A. The electrons are simply those that orbit the atoms of copper in the wiring; this is known as "electron mobility." They move from atom to atom when forced by a voltage induced either by chemical reaction (the battery electrolyte) or by moving magnetic lines of force (the generator).
In order for electrons to flow, there must be a closed circuit. In other words, the electrons must have a path of return from which they started. No electrons are created or destroyed.

Q. Is alternating current (AC) basically the same as direct current (DC) in that electrons move along a conductor? If I were to reverse the leads from a battery to its load 60 times a second, would I have something like AC or would it be simply pulsed DC?
A. Electrical current has two primary elements: electrons moving along a wire very slowly, and a wave impulse that travels through the wire at nearly the speed of light. It makes no difference whether we are talking about AC, DC, pulses, square waves, or sine waves, the electrons do the same thing; only the timing changes.

If you make/break a battery connection on a circuit, you produce pulsating DC, seen on an oscilloscope as square waves. An AC sine wave, however, is produced by a generator gradually changing its electromagnetic coupling between the spinning rotor coil and the stationary field coil. The induced current increases gradually from no coupling to maximum coupling, then down again as the rotor coil passes by the field coil and recedes away, changing polarity.

Q. As a lead-acid battery charges, the current gradually decreases. Is that because the battery's resistance is going up?
A. No, it's because the increasing voltage of the battery actually opposes the voltage of the charger, and once they are the same, no current exchange takes place. A voltage must sense a potential difference in order for it to flow.

If you place a conductor across the terminals of a battery, it senses a potential difference between its positive and negative terminals and current flows because its negative electrons are attracted to the positive terminal. If it sees an identical voltage attached across its terminals, there is no potential difference to cause a current flow, just as when two identical batteries connected in parallel.

Q. I have noticed a "CE" symbol on many electronic products. Is this a certification of some type?
A. Yes, it is French. Conformité Européenne (European Conformity) confirms that the product meets standards set for European consumer distribution. The implication is that the product does what it's supposed to do, and safely.

Q. What is the difference among radio terms like RDS, SDR, and DSP?
A. The Radio Data System (RDS or, in the USA, Radio Broadcast Data System—RBDS) is a means of digitally embedding text in an FM radio station's signal, such as call letters and format, which can be shown on the receiver's LCD display.

Software Defined Radio (SDR) is a classification or design indicating that formerly-analog RF stages like mixers, IFs, and detectors, have been replaced by digital circuitry which can be addressed and manipulated by software commands.

Digital Signal Processing (DSP) is a means of digitally controlling single-signal clarification circuitry like final IF, detector, audio, and noise interference in order to optimize a signal.

Q. Can you identify this family heirloom? It's a 4-pin, 4-1/4" vacuum tube with a line-up pin on its side, and it's marked *Trepassey Azore Plymouth NC May 1919* and the designation GCIG2. It is accompanied by a news photo clipping: *Navy Curtiss NC-4 flying boat, first aircraft to cross the Atlantic 1919.*
A. It was in Trepassey Harbour, between Labrador and Newfoundland, where Amelia Earhart boarded "The Friendship" in 1928 to become the first woman to fly across the Atlantic Ocean. But nearly decade earlier, on May 16, 1919, the U.S. Navy's Curtis Flying Boat, the NC-4, departed Trepassey and flew to Portugal via the Azores, completing the first transatlantic flight; it returned on May 31 to Plymouth, England.

The NC-4 was equipped with in-flight wireless equipment capable of up to 300 miles communications in both voice and telegraph. I suspect the tube is actually a CG-1162 five-watt oscillator, common for military gear of the period.

As to whether this tube was actually used in one of the flights, or for Navy surface support which was extensive for that historic voyage, or just the same type as used in that period is open for conjecture. But it's a valuable artifact, so don't break it!

Q. Whatever became of all the preamps that used to be available for scanners?
A. Scanners improved in sensitivity, negating the need for a preamp. To keep costs down, compromises in design include narrow dynamic range, the ability of a receiver to respond to strong signals without overload.

Wideband preamps aggravate that problem by amplifying a wide swath of spectrum, including very strong signals which cause desensitization and image/intermodulation products.

That said, a preamp can be mounted at an antenna, not at the scanner, to overcome long transmission line losses. .

Q. My scanner no longer has any audio. What can I look for?
A. It could be a bad speaker or earphone jack; plug in an earphone to see if you can hear sound. If not, try the factory reset procedure. Can you still see a digital display, and is it showing that the radio is scanning? If so, does it stop if you rotate the squelch control?

Did you have any advance warning like sound cutting on or off, or did you have to turn the volume or squelch control back and forth because it was intermittent so that it may require an electrical cleaning spray?

If it doesn't pass those tests, it probably needs bench work.

Q. How safe is it to run radios and/or computers from a typical 5kW home generator during power outages, and is the voltage clean enough for this type of application?
A. It should be perfectly safe for the radio equipment. Most commercial gas-driven generators deliver decent voltage stability and waveform; although it will vary a few volts and a few hertz, it's close enough to maintain reliable power to the radio.

Some older, AC-operated (desktop/tower) computer's components, however, may be a little less tolerant of a change in waveform and voltage spikes. This is something you can only determine by experiment, but you won't hurt anything if the generator is operating normally.

If you note changes in the computer's performance while operating it on the generator, power it down and insert a line conditioner between the generator and the computer AC input. A line conditioner is a good idea anytime there is a question about the quality of the voltage and waveform of an AC line. A high-quality battery backup (uninterruptible power supply or UPS) may help as well.

Generally speaking, a laptop or notebook is far less vulnerable to line voltage changes because it doesn't depend on the direct presence of AC; the in-line AC/DC adaptor rectifies the incoming AC to low-voltage DC which charges the internal battery as well as runs the computer. Small changes in the AC voltage and frequency will have no effect on the laptop operation.

Q. I've always adhered to the rule to keep fresh gasoline for my car and for powering an emergency generator. Is it true that gasoline goes bad after a few months?
A. Only in certain cases. Alcohol-supplemented gasoline readily absorbs water vapor from its environment and the gasoline will separate into a top phase (virtually pure gasoline) and a bottom phase consisting of water and alcohol, making the engine difficult or impossible to start.

In a two-cycle engine utilizing a gas/oil mix, the engine may start with just the bottom water/alcohol mix that is devoid of the lubricating oil that is in the top phase with the gasoline, thus damaging the engine.

If there is any concern about storing gas for several months, such as for an emergency power generator, simply add a fuel stabilizer available from auto supply stores, hardware stores, and the automotive sections of department stores. To minimize water pickup, keep the tank full and tightly capped.

Q. Does the FCC ever go after those CB operators using linear amplifiers?
A. Enforcement of the provisions set by the FCC regarding the CB radio service was essentially abandoned by the FCC some years ago, due to lack of manpower under the enormous volume of complaints from irate citizens.

A Congressional Bill was signed into law by President Clinton November 29, 2000, permitting state and local governments to set enforcement regulations for violations of FCC rules and regulations of the CB radio service, but it's rarely implemented.

Q. My cell phone battery charger will charge for a period of time, then shut off. If I remove the battery, then plug it back in, it charges again briefly. Is this normal?

A. It sounds as if your cell phone charger is doing its job properly. There are three common types of chargers:

 1. The trickle charger which delivers a low current charge for 12-16 hours;

 2. The voltage-sensing charger which periodically samples the battery's state of charge ("terminal voltage") before it shuts down; and

 3. The fast charger which delivers a relatively high current rate for a short time.

Q. In WWII movies, we often hear the radio operator respond to a command, "Roger wilco." What is the derivation of this, and why do hams say "73s"?

A. "Roger" is the phonetic for the letter R, a hold-over from the Morse code abbreviation for "Received." "Wilco" is the voice contraction for "Will comply."

 In the early days of radio, a numeric shorthand system was developed for speedier commercial and amateur telecommunications. 73 meant "best regards." Today, many ignore the fact that it's already pluralized and say "73s" which would literally translate to "best regardses."

Q. What are the parameters which must be met before UL approval must be sought? I never see UL on car switches or fuses.

A. Underwriters Laboratory is a private corporation that certifies the safety of electrical products on a client-by-client basis. It has no legal status, but their reputation is held in high regard, so many companies require the UL certification when they buy products.

Q. In regard to the proposal to put a tracking chip in firearms so they can be located, is this technically feasible?

A. Yes, theoretically. These are available to track diplomats, children, pets, shipments, and migratory wildlife. The chip could be hidden in the non-metallic stock or grip, but store the gun in a metal case and it's all over; that's a Faraday shield, outside of which no signal would be detected.

Q. According to SETI (Search for Extra Terrestrial Intelligence), we have been sending signals out to the surrounding universe for some 70 years. Wouldn't this just be white noise to an alien?
A. The signals vary in content, some with video and audio, others as photos, text strings, or even mathematical or geometric progression. It is assumed (hoped) that if heard by ears or seen by eyes not familiar with our languages or music, the mathematical repetition, progressions, and other unnatural relationships should indicate their artificial origin to a distant listener, and encourage decoding or demodulation.

Q. The alternator on my truck puts out 13.8 volts. Does this mean that the battery is trickle charging all the time?
A. Yes, until the battery's terminal potential builds up to 13.8 volts. Then its own potential opposes the charger voltage so there is no current flow.

Q. My old Ford Mustang had an ammeter in which a needle pointed to either charge or discharge. Is this better than a voltmeter?
A. An ammeter reveals how much current is being drawn, but that can happen with a good or bad battery, or a partially-discharged good battery, so it doesn't diagnose the battery's condition.
A voltmeter, on the other hand, will show the terminal voltage when the car isn't running, indicating whether or not the battery is holding a charge. It will also reveal how badly that voltage drops when it's trying to start the engine, revealing the battery's condition. I'd go with the voltmeter.

Q. Is there any difference between speaker wire and regular zip cord (AC lamp cord) other than the clear insulation and (I think) lower price? Is this simply marketing?
A. Yes, marketing and some legally. 120 VAC zip cord (lamp cord) has to meet crucial safety standards to avoid fire and electrocution, while speaker wire doesn't since audio voltages are so low. From an audio standpoint, zip cord makes great speaker wire, and you don't need

heavy-duty monster cable unless you're running thousands of watts into hundreds of feet of wire!

Q. What causes the bluish-white corrosion on a car battery's positive terminal?
A. The battery is filled with sulfuric acid which is actually a solution of hydrogen ions (positively charged molecules) and sulfate ions (negatively charged molecules). Seepage of this battery acid coats the surface between the battery posts, conducting a minute electrical current between those terminals.

Since opposite charges attract, the negative sulfate ions migrate to the positive terminal. Here they combine with the lead to form lead sulfate (the white fluff), and with the copper wire to form copper sulfate (the bluish fluff). The hydrogen molecules simply evaporate, that's why we don't see any deposit on the negative terminal.

Q. My new floor lamp takes standard light bulbs, 60 watts maximum. But the instructions say not to use cold cathode fluorescent lamps (CFLs) exceeding 13 watts, and I'd like to use my 30 watt CFLs. What gives?
A. There's nothing wrong with putting in those 30 watt CFLs; they will operate cooler than the 60 watt incandescents, and provide brighter light as well!

Q. I was doing some electrical work and I connected a multimeter to the hot wire and then to a nearby ground rod. I got a 120 volt reading. Does this mean that the current returned to the nearby step-down transformer through the ground rod attached to our meter base neutral as well as the ground wire on the transformer?)
A. Yes. In all regulation, three-wire grounding systems, one wire goes from the "neutral" side of the transformer to the wide flat pin of the wall socket. It also is earth-grounded, as is the round pin of the wall socket. The hot wire will read 120 VAC touching either of these.

Q. I have heard that NASA keeps an eye on every piece of "space junk" that orbits the earth. Is this true and, if so, how do they do it?
A. Orbital space debris, which includes everything from particles and paint chips to satellites and space stations, numbers in the tens of millions. Keeping track of all this is not NASA, nor is it NORAD. That job falls to the Joint Space Operations Center of the U.S. Strategic Command at Vandenberg Air Force Base, through the Space Surveillance Network (SSM), a global network of 30 space surveillance sensors that include military and civilian radar and optical telescopes used to observe these objects. SSN makes up to 420,000 observations each day.

16,000 orbiting pieces are currently being tracked. Smaller particles are considered relatively harmless against the rubber bumpers on the International Space Station (ISS). These tiny particles travel at a velocity of up to 17,500 miles per hour—more than 20 times the speed of sound! Depending on size, the impacts can be consequential.

Q. If I buy a four-port splitter for future scanners and only use two of the ports now, do I need to terminate the unused ports?
A. No, it's not necessary, and ignoring them may actually provide higher signal levels to the two scanners, since terminating resistors would properly match the two unused ports, guaranteeing that half of the signal voltage is dissipated there.

Q. Is there a reliable way to disable an imbedded RFID tag?
A. A microwave oven will destroy a Radio Frequency IDentification tag, but could also damage what it's in, like a fabric which might catch fire. An RFID tag can also be disabled by simply wrapping or shielding it with metal foil, or putting it inside a metal enclosure, either of which acts like a Faraday cage.

The maximum range of a passive RFID tag is just a few feet since it has no power of its own and is powered by the reading device. Active RFID tags contain batteries, allowing them to radiate a signal much farther.

Q. I have an old scanner that has developed a hum through the speaker; what is the likely cause?
A. If the hum stays the same level with the volume turned down, it is likely to be either a bad filter capacitor in the power supply, or a shorted rectifier diode, also in the power supply (less likely). Filter caps contain an electrolytic chemical and are subject to drying out with time, thus losing their filtering ability. If the radio functions normally except for the hum, it's the capacitor, not the diode.

Q. I recently placed a wire across a magnetic compass and passed 10 amps of AC current through it; the needle barely budged. Then I passed 10 amps of DC current through it, and the needle swung in line with the wire. Why the difference?
A. When electrical current passes through a conductor, it produces a magnetic field. Just so long as the current is passing in one direction and remains constant, the magnetic field is uniform and attracts the magnetic needle of the compass.

But AC fluctuates back and forth, reversing its polarity 120 times per second. The needle may vibrate, but can't deflect long enough in one direction before the field reverses, pulling it back in the other direction, so it appears to sit in one position. A rough analogy would be two people rapidly and alternately pushing, and then relaxing, on opposite ends of a car—it wouldn't ever get rolling.

Q. Would my shortwave portable and handy-talkie be in danger if I leave the batteries in them and power them up now and then?
A. Generally speaking, the only thing that could blow up a battery would be high heat from excessive current charging, or shorting it out (which is unlikely to occur in a battery compartment). If the battery never gets hot (warm is OK), this should not be a problem.

However, not all batteries are immune from chemical leakage of their electrolyte, especially when discharged. Caustic chemical electrolytes can damage a battery compartment, its contacts and nearby circuit board copper traces.

It's always best to remove batteries (rechargeable as well as non-rechargeable) if the equipment is left unused more than a few weeks.

Q. What causes the musty smell that comes from an electric heating system when it's turned on after being off for the season?
A. The familiar odor is caused by colonies of fungus, algae, and molds which like their cool, dark quarters in the summer until you cook 'em in the fall.

Q. Are there any electrical medical devices that provide pain relief based on the Tesla coil? Surely there must have been some positive results as the man was truly a genius and ahead of his times.
A. There's no question that Nicola Tesla was a genius, way ahead of his time in terms of practical applications for his alternating-current (AC), high-voltage transformers, but they didn't have any validity in treating pain.

Typical pseudo-medical applications of Tesla's high voltage were the "violet ray" devices which had a hand-held glass wand wired to a step-up transformer. The name "violet ray" referred to the blue glow emitted from the partially-evacuated glass electrodes inserted into the end of the wand, the high voltage ionizing (electrically charging) the remaining air, mostly nitrogen which glows blue under those conditions.

An earlier device, the Faradic battery, had a lower voltage output and was applied directly to the body rather than indirectly through an ionized-gas, glass electrode.

Both schemes had intensity (voltage) controls adjustable to different discomfort levels. You can find many of these devices on eBay by searching for "quack medical," "violet ray," and the like.

Currently, the more modern transcutaneous electrical nerve stimulation (TENS) units produce alternating current in the 1-260 Hz range which reportedly blocks pain.

Q. What is the name of the flexible pin plug that can be inserted into the center of a standard shortwave/CB-style female antenna connector (SO-239) so a single wire can be attached?
A. Because of its shape, it's known as a banana plug. They are also commonly used on test prods for multimeters as well.

Q. I'm a newcomer to shortwave radio. Would you mind answering a few questions for me?
A. Many radio hobbyists are newcomers, and they would benefit as well from your excellent questions:

(1) Is it true that current amateur radio licensing does not require Morse code?

Yes. Since 2007, no knowledge of the Morse code is needed to pass an amateur radio exam.

(2) What is the difference between shortwave radio and amateur radio?

Shortwave radio is very general reference to all users of the high frequency (HF) spectrum (3-30 MHz). Above that are very high frequency (VHF) from 30-300 MHz, and ultra high frequency (UHF) (from 300-3000 MHz). The microwave services are even higher in frequency.

Amateur ("ham") radio, on the other hand, is a specific licensed radio service which has FCC-specified frequency bands found throughout the radio spectrum, from the low frequency (LF) range well into microwave.

(3) What are the advantages/disadvantages of shortwave radio?

Shortwave frequencies carry signals quite far; the signals may propagate over mountain ranges; inexpensive radios to hear them are readily available; and simple wire antennas are all that are necessary to receive them.

The shortwave spectrum, however, is vulnerable to severe electrical interference from local power lines and electrical appliances; noise from electrical storms over considerable distances; severe signal attenuation during low sunspot cycles; radio interference from distant stations on the same frequencies ("skip"); and blackouts from solar flares.

(4) What types of stores sell shortwave radios?

Amateur radio stores can be found online and advertise in amateur and hobby radio magazines like CQ and The Spectrum Monitor.

(5) Are there programs on shortwave, as there are programs on regular radio, or do people just talk back and forth?

There are both. Many two-way "utility" users of the shortwave (and higher) radio spectrum, include federal government, military, airlines, maritime services, Coast Guard, and more.

(6) If there are programs, how does one obtain permission to have a program, and what determines who and how far away the program will be heard?

Licensing is required by the government of the host country; in the U.S., it's the FCC. Most shortwave broadcasters are either evangelical, using the airwaves to propagate their religious views, or government sponsored, propagating their political views. There are also entertainment, news, educational, and music broadcasts.

Q: Is it possible that the weakening of the earth's magnetic field over geologic time, including pole reversals, could be affected by man's artificial magnetism?

A: No. Scientists believe that the earth's magnetic field is generated by circulating electrical current within its iron-nickel core, a "dynamo" effect. While the field is weak, even the total magnetic energy of all artificially generated electromagnetic fields created by technology pale in comparison.

Man-made magnetic fields are confined, erratic, and isolated when compared to the giant magnetic earth. And the electromagnetic fields we generate, from 60 Hz on up through microwaves, scarcely penetrate the earth's topsoil.

Q. My neighbor is planning to install an electric fence for livestock control. Can these be fatal? What is the liability for electrocution?

A. A high-powered fence charge may put out as much as 8000 volts DC, but it's at very low current (1 or 2 milliamps). Lethal current is at least 6 mA, so the fence may not be lethal, but it sure can sting!

I don't give legal advice, but it would be wise to prominently mark the fence at intervals to minimize liability from negligence.

Q. After accidentally leaving my car headlights on too long, the battery died. I tried jump-starting with a freshly-charged battery, but with the dead battery still connected—or even disconnected—the engine barely turned. Do jumper cables really add that much resistance?
A. Yep, you guessed it. The starter motor on a vehicle takes quite a current—several hundred amps—to crank. Let's do the math with Ohm's law: $E=IR$ (volts = amps x ohms).

If our theoretical 12 volt battery must deliver 300 amps cranking current, that means that the total resistance of the auto's starter motor and cabling is only 0.04 ohms! If those long jumper cables have even a fraction of an ohm, the drop from the booster battery is considerable. I just measured the total resistance of the two conductors on my own copper-wire jumper cables: 1 ohm!

That's why a jump start turns the motor so slowly, and the battery cables in your car are so short and thick. It's always a good idea to let the jumper battery charge the car battery for a few minutes so both batteries can try starting the engine.

It's also helpful to have the engine running in the car that has the booster battery so its alternator (generator) can add its power to the jumper system.

Q. Is there a formula to convert a lead-acid battery rated in cold cranking amps (CCA) to ampere hours (AH)?
A. Unfortunately, no. Batteries used for starting automotive engines have more lead and are capable of providing heavy current loads for a short period of time as they turn the engine over, while electronics batteries have less lead and provide moderate current loads for a longer period before the voltage drops appreciably.

For example, a car battery may be rated for 500 CCA, meaning it can provide 500 amps at 0 degrees Fahrenheit for 30 seconds before the terminal voltage drops to 10.8 V, while a 7.5 AH computer backup battery can provide 7.5 amps for an hour before it discharges to 10.8 V (below which a lead-acid battery can experience damage).

Because the two applications require different battery construction, I have seen references that show anywhere from 3 to 10 as divisors to convert CCA to AH; thus, a *very* approximate gauge is to divide CCA by 6 to get AH. While you won't get an accurate number, it can let you know if you're in the ball park!

A much better way is to connect a 12V headlight or brake light bulb on a fully charged battery and see how long it takes to drop to 10.8 V while monitoring the current. Keep in mind, however, that as the voltage drops, so will the current through the bulb, and the bulb's resistance.

For a more accurate measurement, you should keep changing the resistance to maintain a constant current drain while monitoring the voltage as it reduces to 10.8 V.

Q. I have an excellent external speaker I'd like to add to my receiver, but it's rated at 4 ohms impedance. My radio specifications say to use an external speaker of 6-8 ohms. What problems might I encounter?
A. While a solid-state audio amplifier is designed for a certain power maximum (watts), it's the current (amps) that heats them up, sometimes destructively high.

Following Ohm's law ($W = I^2 R$), we see that for the same amount of power to be produced in a 4 ohm speaker as in an 8 ohm speaker, four times as much current must be delivered.

Additionally, it takes twice as much voltage to produce that extra current, and high voltage is another problem for solid-state devices. Even higher voltages are generated by the impedance mismatch similar to VSWR on radio transmission lines.

You probably won't do a bit of harm using the 4 ohm speaker just so long as you don't keep it at high volume. A failsafe technique to prevent this from happening is to add in series a low-value resistor (2.7

ohms at 1/2 watt) to keep the total impedance at or above the specified 6 ohm minimum.

Feel the resistor after a few minutes of reasonably loud listening to be sure it doesn't get uncomfortably hot. If it does, try a higher power (1 watt) resistor or, better yet, turn the volume down.

Q. What makes digital TV signals freeze and drop out with weather conditions rather than simply fade?
A. While our traditional analog signals simply competed with analog background noise ("snowy" picture), modern digital transmissions are all or none. When the data stream is interrupted or corrupted by even low levels of interference, the entire data stream drops out.

Q. The push-to-talk switch on most transceiver microphones disconnects the microphone element during the receive function. Why is that necessary?
A. Since the same audio circuitry is active for both transmit and receive, the mike being on during the receive function would produce acoustic feedback from the speaker.

CBers who enjoy the echo sound of feedback can acquire models that actually allow leave the speaker connected for that sound effect.

Q. Why is AC more efficient in long-distance power lines than DC?
A. It isn't more efficient, it's just easier to convert (transform) to other voltages.

AC can be directly transformed to any voltage by a simple turns ratio, while DC must either be dropped in voltage by resistance (very lossy as heat), or fed into a DC/AC converter to step it up. That is not as efficient, and is more complex and expensive than using a simple transformer.

Q. What is the basis for radio rack panel height (multiples of 1.75") and width (19") standards?)

A. The original standard was developed around 1890 by George Westinghouse for the railroads to use as a mounting system for railroad signaling relays, thus the common name, relay rack. He later adapted it to the telephone industry to mount their array of relays. Eventually, the Electronic Industries Association (EIA) adopted the system.

Q. I enjoy monitoring aircraft communications. When someone says, "flight level 380," does that mean that the aircraft is flying at 38000 feet? How high can a commercial airliner fly safely?
A. Yes, FL380 is a flight level of 38,000 feet. It's an abbreviated response, just like when the tower replies, "Contact (airport name) 121.62" they really mean 121.625 MHz.

Higher-performance aircraft can cruise as high as 51,000 feet (nearly10 miles), with one record set for the Concorde at 60,000 feet. Lowest flight levels are more tightly restricted as a safety precaution because of the number of aircraft cruising those altitudes.

Q. I'm reluctant to replace my filament light bulbs with expensive compact fluorescent lights (CFLs). Is there really a savings?
A. Traditional tungsten-filament bulbs waste 90% of their electricity as heat; only 10% of the energy consumed is emitted usefully as light. The new CFLs use 75% less electricity to provide the same amount of light as traditional incandescent bulbs, and last ten times longer. LED bulbs use 85% less electricity for the same illumination, and last forty times longer.

Prices continue to drop. I suspect that CFL bulbs will eventually give way to LED for the long term. LEDs respond instantly with full brilliance, produce no radio interference as some CFLs do, are much more durable, and contain no toxic substances like the solder in incandescents and the mercury in CFLs.

Q. How do U.S. AM radio stations continue to operate when Canada and Mexico are abandoning the medium wave band?

A. With economic unrest and political upheaval, talk radio is a current rage on AM in the U.S. As long as there is an audience, there will be sponsors, and as long as there are sponsors, there will be a profit.

Q. I don't trust a car battery with more than five years on it. A three-year-old battery once failed on me without warning. I recently purchased a 100 amp load tester. If the tester indicates a good battery, can I trust it regardless of the age?
A. Over time, lead-acid storage batteries build up sulfate deposits on the lead plates. Batteries should be continuously "exercised" (charge>discharge> charge>discharge, etc). The best recharging seems to be with pulse-type chargers, and the worst is having a trickle charger always on.

Five years with a well-maintained vehicle battery is possible, and three years is likely. Your 100-amp load tester should dependably indicate the state of a battery since sulfation is a slow process; the load tester should reveal when it's on its way out.

Q. How does carbon-14 dating work, and how accurate is it?)
A. Constant bombardment of our atmosphere by cosmic rays converts a small amount of the nitrogen into carbon 14 (^{14}C), an unstable radioactive form of carbon that gradually "decays" over time back to a stable nitrogen atom.

Plants and animals absorb the ^{14}C during normal respiration, and when they die, the ^{14}C starts diminishing. The rate of decay is known as the half life, the period over which half of the ^{14}C changes back to nitrogen.

When that time period repeats, half of the remaining ^{14}C decays, then half again, and so on. The half life of ^{14}C has been established at 5730 +/- 40 years, and remains stable until about 60,000 years when the small numbers of remaining radioactivity approach randomness.

Q. What is a Q meter and what is it used for?
A. "Q," shorthand for "quality factor," It refers to the ability of a component like a coil or capacitor to store the energy induced by a

signal without introducing resistive losses. It is a ratio, so it has no unit.

A high-Q component would process a signal predictably as its lossless design would expect, while a low-Q component like a small coil of fine, poorly insulated wire dobbed with moisture-laden potting compound would introduce consequential resistive losses, especially at higher frequencies.

Q. Can I use my old NiCd charger to recharge newer NiMH batteries?
A. An automatically timed NiCd charger will under-charge a NiMH because the latter has higher capacity, so the timer is likely to turn off before fully charging the newer chemistry.

If it's one of those low-current "overnight" chargers, then it will take much longer to charge the NiMH, perhaps a couple of days. If it's a rapid charger, then check periodically to make sure it's not overheating the battery. Warm is fine; hot is not.

Q. Why is so much fiber being used in place of cable for data transmission?
A. Fiber is inherently much faster, allows wider bandwidth capacity, is lighter in weight, has a smaller diameter, doesn't corrode like copper, and doesn't suffer insulation signal losses.

Q. I'm trying to power a portable TV that uses 5 VDC. I have a transformer that will give me 6 VDC. Will there be any damage from the extra volt or should I spring for the real thing?
A. It's highly unlikely that you would cause any damage by using the 6-volt power supply. Just monitor the temperature of the little TV; if it gets uncomfortably warm after just a few minutes, then I'd say it's overpowered. If the TV isn't overly warm and the picture is just fine, have at it!

Q. Do photocells ever wear out?
A. Yes. Manufacturers give them a nominal 25+ year warranty, with gradual decay approaching 40 years.

Modern solar panels live longer than they used to, but still age faster in their first couple of years. After that, depending upon their chemistry, they degrade less than 1% per year. Currently, monocrystalline silicon (mono-Si) is considered best (0.36% per year).

Q. Do so-called "silent" dog whistles actually work? What pitch is their ultrasound?
A. Yes; I've tested one on my collies. It's made like a piston with an adjustable screw; as you tighten the screw, it shortens the chamber, thus raising the pitch. Most adults choose a frequency approaching 20 kHz for it to be inaudible to them.

Q. I have been looking for a hand-held ham rig in the event of a disaster. I'd like a radio that will reach at least across town. I know I need my license, but I'd like to get the rig first as an incentive to take the test, but really don't know where to start.
A. I'd recommend a two-meter (144-148 MHz) handy-talkie. These are widely available and inexpensive. Depending on the model, they can cost under $50 new, and you can talk simplex (direct radio to radio), or through a repeater in your area for broader coverage.
    The Technician Class license is easy to get and there's no longer a Morse code requirement at any level. For more information, visit the American Radio Relay League (ARRL) site at: http://www.arrl.org/licensing-education-training.

Q. I read about security compromise of the RFID chips which are presently equipped in US passports. For those of us holding US passports with RFID chips, can you recommend any countermeasures we could take to preempt such a compromise?
A. The unwanted intrusion can be virtually eliminated by metal shielding, either wrapping the passport in aluminum foil, or perhaps simply carrying it in an anti-static bag. The higher the frequency used, the easier it is to shield from the interrogation system. In radio parlance, this is known as a Faraday shield.

Q. Some hobbyists leave their radio equipment on continuously rather than shutting it off when in disuse. Is there any advantage one way or the other?
A. Probably not. In the old days of vacuum tubes, there were heat and high voltage issues, but modern, solid-state equipment doesn't suffer from that. Assuming the equipment does not get hot, the only considerations would be unnecessary power consumption from the grid (both economic and environmental concerns), and internal cooling fans constantly drawing in dust.

Q. A friend of mine has a flashlight that recharges itself by shaking it. What's inside to make it work?
A. The plastic barrel of the flashlight is internally wound with a coil of wire. When the flashlight is shaken, a strong magnet slides back and forth, inducing current into the coil. That current is used to charge a battery. Essentially, the flashlight is a crude magneto form of generator like the old, crank-telephone ringers.

Q. What is Litz wire and why was it used in early radio?
A. Radio frequency (RF) currents have a tendency to travel near the surface of a wire, not all through it; hollow wire would work just as well. The higher the frequency, the more the RF currents migrate toward the wire's "skin."
    Litz wire is a woven cluster of fine wires separated from each other by strands of cloth insulation. As a result, the RF currents have several conductors for them to travel near the surface, thus reducing the resistance they would encounter with only one conductor.

Q. When transmitting roughly 100 watts on the 17 and 20 meter ham bands, my carbon monoxide alarm goes off. What causes this?
A. When RF voltages are impinged on wiring, the wires behave like antennas, delivering voltage to parts of the affected equipment that are sensitive to any change in voltage levels, such as detectors of various sorts. I suspect your CO detector interprets the increase in electrical

voltage the same as if its sensor was sending the voltage, thus triggering the alarm.

Q. Growing up with radio, I survived the change from megacycles to megahertz. I learned that resistance is measured in ohms, kilohms (x1000) and megohms (x1,000,000), while capacitance was in microfarads and picofarads; so what the heck is a nanofarad?

A. As in your example, most scientific units of measurement are conveniently classified in multiplier intervals of 1000, just like milligrams, grams and kilograms.

Capacitance is usually measured in microfarads and picofarads, separated by a multiplier of 1,000,000: 1 microfarad = 1,000,000 picofarads, a shift of six decimal places.

To conform to the standard classification scale and thus avoid awkward decimal values, the nanofarad is sometimes used. Thus, a 0.001 microfarad (1000 picofarads) capacitor is also 1 nanofarad – simply a shift of the decimal three places. But cheer up, the use of microfarads and picofarads is still far more common.

Q. It seems like the next big fad could be "wireless electricity" in which power is radiated into a home or office, and picked up by sensors rather than conducted by wire. Is this an old idea whose time has come, or are we in for yet another source of wideband noise once this idea gains significant market share?

A. Unless the laws of physics change substantially, I don't see this mode of power transmission coming in the foreseeable future. The problem is not one of interference, but inefficiency.

While the promoters say that they get upward of 90% efficiency when an inductively-coupled accessory like a toothbrush is sitting right on the charger, the marketers envision the invisible transmission of power throughout buildings to power lamps, computers, TVs, and other heavier duty accessories.

At such separation, the efficiency drops dramatically. When you hear a distant base station on your scanner, it's the results of hundreds of watts in the transmitting antenna dwindling to millionths of a watt by the time the signal reaches you.

Wrapping your home in wire coils so that you and your equipment are bathed in the center of an electromagnetic field invites serious health concerns. Electrical cords in the home and office are still the answer for much time to come.

Q. Can a spark occur in a complete vacuum? Would you see or hear it? Would more or less voltage be required to produce it?
A. It can occur, but you wouldn't see it because the blue light is produced by the fluorescence of the ionized nitrogen in air, but if there's no gas present, there's no ionization and no light. Nor is sound produced by the rapid expansion of air which is missing.

If you do hear a sound when a spark occurs in a vacuum, it would be produced by mechanical movement in the current-carrying wiring outside the vacuum chamber.

Electrons will still flow, but more voltage would be needed to produce the spark, since even the high resistance of air is more conductive than no air at all.

Q. If I insert one test prod from my digital voltmeter into the hot pin of an AC wall socket and let the other dangle, I get a reading of several volts. If I touched the sdangling prod with my fingers, the reading is higher. Why is this, and what would happen if I attach the dangling prod to a large metal surface like a car body?
A. You were measuring the electrical potential between the hot wire of the AC line and whatever minute return there could have been between the resistive, humid air and common ground for the AC system.

When you held the second prod with your fingers, or connected one prod to a car body, the area of your skin and the car body coupled more electrical conductivity into the system.

A simple analog AC panel meter would have shown a lower reading because its lower resistance would have caused a voltage drop by demanding more current to flow to activate the meter; modern digital voltmeters (DVMs) don't require much current to respond with their reading.

Q. What is the difference between the earth's magnetic poles and geomagnetic poles? Why do the poles wander over time?
A. The magnetic poles are the north and south points on the earth's surface where a compass needle would point down to the magnetic poles of the earth's core. The geomagnetic poles are the distant north and south regions where a surface compass points because the core's magnetic lines of force emerge from there.

The poles wander several miles per year as the molten magnetic core material circulates.

Q. I heard that if you drive a couple rods a foot or so into wet soil and apply house current to them, worms will start coming to the surface. Is there any truth to that?
A. There certainly is; it's an age-old fisherman's trick more commonly used in the past with magnetos from hand-crank telephones. But there's a delicate balance between stimulating worms and electrocuting them. Much depends on the moisture and mineral conductivity of the soil, and the length and separation of the rods.

Q. While tuning through the shortwave spectrum, I used to hear voice transmissions reading lists of numbers. What are these, and are there any books written about them?
A. There's plenty of information on the web about these spy numbers stations sending routine messages to agents in foreign countries—Cubans in the U.S., Americans in Cuba, etc.

They start with a "header" that addresses a particular agent, then the code is sent which can be decoded only with a one-time pad, a pack of pages which changes every day to thwart deciphering by code breakers.

At one time, the strong, Spanish-language signals were from an old, WWII, German-made broadcast transmitter, and originated from just outside of Havana. The English-language transmissions came from a U.S. Army communications installation in Remington, Virginia near the Warrenton Training Center.

Q. Just what is geothermal energy and how is it used?
A. The deeper you dig in the earth, the warmer it gets. Most of the geothermal energy is used for heat pumps, providing space heating to homes, businesses, and factories. Some of the hot gases and water are used to convert the heat into electric power by driving generators.

At present, the majority of geothermal pumps are located on the edges of tectonic plates (cracks in the earth's crust) where there are volcanoes, geysers and hot springs because hot liquids and gases are more readily reachable from the earth's surface. Newer technologies permit efficient energy production at lower temperatures, allowing wider geographical distribution of these systems.

Q. I recall seeing automotive batteries in photos of Hussein's torture chambers and in fiction movies as well. Is 12 volts enough to cause pain?
A. The damage and pain produced by electricity depends on the current flowing through the body: 1 mA can be felt; 5 mA is painful; above 15 mA a person loses muscle control; 70 mA can be fatal. The higher the skin resistance, the less current will flow.

I tried wetting both my hands, then measured the electrical resistance with an ohmmeter while touching two large metal electrodes. My wet skin resistance, hand to hand, was roughly 10,000 ohms, so the current flow from a 12 volt car battery would have been about 1 mA.

Locating the contact points closer together or puncturing the skin would allow much lower resistance, and connecting several batteries in series would increase the voltage.

Q. Wouldn't it be good to roll up loose cables (AC cords, coax, audio lines, etc.) in radio rooms into coils to act as chokes to reduce interference in their attached equipment?

A. That precaution wouldn't do much at the lower frequencies, just for VHF/UHF where it's not as much of a problem. Better, use ferrite bead chokes on all such leads, placed close to the equipment cabinets.

Q. As happens to everyone, I occasionally get shocked with static electricity; I can even see a small spark. Roughly how much voltage are we talking about here?

A. When dielectrics (non-conductors like plastic and fur) are rubbed together, considerable static voltage can build up. A general rule of thumb is 30,000-70,000 volts per inch of spark depending on air pressure and humidity, so you probably are getting zapped with a few thousand volts! Fortunately, the current is so minute that no damage is done.

Q. On a recent road trip, as I set my scanner down on a desk in my motel room, I remembered that some disreputable motel/hotel managers install two-way mirrors in the wall to peep on the clients. How can I tell if my room has one of these?

A. All commercial mirrors are back-surface silvered; when you put your finger tip on the surface, you will see a slight gap between the tip and its reflection. If it's a two-way glass, there will be no gap because you are relying on the front surface for a reflection.

Two-way glass is only about 12% reflective, and mostly transparent, so you can also see through them. With the lights out in your room, look for hints of light behind it; if you press a flashlight against the mirror, you will see its beam on a wall behind it.

Finally, mirrors are installed against an existing wall and may, at most, have a simple frame around them. If you tap on one, it sounds solid. A two-way glass is mounted within the wall in a substantial frame; if you tap on it you will hear that deeper, hollow, "bonk" sound like a window.

Some unsavory motel/hotel operators cut a small peephole through the wall and scrape a tiny dot of silver off to look through. If that's the case, you will see a dark spot in the silver.

Q. Are the little battery chargers that claim to revive and recharge all types of batteries – alkaline, NiCad, NiMH, etc. -- really effective, or are they just a scam?
A. Different battery chemistries require different charging methods. While it's true that simply hooking an external DC source to a discharged battery will charge it somewhat, the method doesn't fully charge every kind of battery. Some require a constant voltage and some require a constant current.

Alkaline battery chemistry is destructive; once the chemicals have been used up in their internal reaction, they can't be fully recharged, and only partially recharged a few times before they are dead as a doornail.

I'd say that this charges-all-batteries device is in the same class as the glue-on strip that claims to increase cell phone range, the plastic ball to put on the top of your mobile whip, and wrapping your TV rabbit ears with aluminum foil.

Q. I can hear the crackle from distant lightning strokes on my shortwave receiver, but not on my scanner. Is this because the scanner receives FM and the lightning strokes produce an AM signal?
A. Yes and no. Lightning does produce an instantaneous AM signal, rich in harmonics which dissipate the higher in frequency you go. Even FM receivers have some AM detection, as evidenced by images of aircraft signals occasionally heard in the VHF-FM public safety bands.

Lightning can be heard for hundreds of miles down in the low frequencies, and tens of miles in the shortwave bands, but by the time you are in the VHF spectrum, if you hear the lightning on a scanner, you'd better duck—it's close!

As a side note, on a clear day, if you have your shortwave receiver hooked to an outdoor antenna, tune to a dead spot so you don't even hear electrical noise, and then disconnect the antenna; you'll note that the background hiss diminishes.

That vanished hiss is ionospheric "white noise" or "sferics," the combined lightning-stroke energy radiated by over 2000 worldwide electrical storms going on at any one time, producing 40 flashes per second!

Q. When I'm out making the rounds of thrift shops looking for electronic bargains, I often see stereo speakers. Is there a simple test that I can make to get an idea of whether they will provide decent sound?

A. Since it's unlikely that you will be carrying a sweep-tone generator with you, a simple test will give a valid indication of whether or not the speaker is working. If the speaker cone is just a few inches in diameter, it will probably serve just fine for voice, Morse, and data reception; a larger speaker in a larger cabinet provides the bass for music.

Briefly touch a nine-volt battery across the speaker terminals (it won't harm the speaker if done quickly). If the sound is crisp and raspy, it should work for those first-mentioned modes; if it provides a good, bassy "thump" as well, it should work well with music.

If the speaker is enclosed in a wood cabinet behind a grill, you should be able to pull the grill off; it is often on a separate frame with small plugs which detach from the cabinet. This is usually revealed by lightly prying the edges of the grill to observe movement.

Inspect the paper cone to be sure it isn't torn, and that the rubber surround which attaches the cone to the frame isn't crumbling and disintegrating, very common on thrift-shop speakers. While a minor paper tear on the cone can often be repaired with tape, rubber glue or contact cement, the rubber surround can't.

If the cone and surround look good, press gently and simultaneously both sides of the cone and listen for it to rub the magnet; it should move without scraping. Those easy tests should do the trick.

This concludes the list of pertinent questions from readers gleaned from the final ten years of Monitoring Times.

FREQUENCY ALLOCATIONS: THE FIRST 1000 MHZ

FREQ. MHz MODE ALLOCATIONS (PRIMARY secondary)

FREQ. MHz	MODE	ALLOCATIONS (PRIMARY secondary)
0.009-0.189	USB	COASTAL UTILITIES (0.1357-0.1378 AMATEUR CW)
0.190-0.435	AM	NAVIGATIONAL BEACONS
0.436-0.529	USB	COASTAL UTILITIES
0.530-1.700	AM	BROADCASTING
1.701-1.999	LSB	BEACONS (1.8-2.0 AMATEUR)
2.000-2.499	USB	MARITIME
2.501-3.199	USB	UTILITIES (2.850-3.000 Aeronautical; 3.026-3.152 Mil. Air)
3.200-3.400	AM	BROADCASTING
3.401-3.499	USB	UTILITIES (Aeronautical)
3.500-4.000	LSB	AMATEUR
4.001-4.999	USB	UTILITIES (4.650-4.700 Aeronautical; 4.700-4.750 Mil. Air)
5.000	AM	WWV Standard time broadcast
5.001-5.899	USB	UTILITIES (5.450-5.680 Aero; 5.680-5.730 Mil. air; 5.332-5.405 Amateur)
5.900-6.210	AM	BROADCASTING
6.211-7.000	USB	UTILITIES (6.525-6.580 Aeronautical, 6.580-6.765 Mil. Air)
7.001-7.099	LSB	UTILITIES
7.100-7.550	USB	UTILITIES (8.815-8.965 Aeronautical; 8.965-9.040 Mil. Air)
9.39-9.970	AM	BROADCASTING
9.971-9.999	USB	UTILITIES
10.000	AM	WWV Standard time broadcast
10.001-11.579	USB	UTILITIES (10.050-10.100 Aero;10.1-10.150 Amateur; 11.175-11.400 Mil. air)
11.580-12.110	AM	BROADCASTING
12.111-13.599	USB	UTILITIES (13.200-13.260 Mil. air)
13.600-13.845	AM	BROADCASTING
13.846-14.669	USB	UTILITIES (14.0-14.35 Amateur)
14.670	AM	CHU Standard time broadcast
14.671-14.999	USB	UTILITIES
15.000	AM	WWV (Standard time broadcast)
15.001-15.099	USB	UTILITIES (15.010-15.100 Mil. air)
15.100-15.710	AM	BROADCASTING
15.711-17.554	USB	UTILITIES
17.555-17.900	AM	BROADCASTING
17.901-19.999	USB	UTILITIES (18.068-18.168 Amateur; 17.900-17.970 Aero; 17.970-18.030 l. air)
20.000	AM	WWV (Standard time broadcast)
20.001-21.449	USB	UTILITIES (21.0-21.450 Amateur)

21.450-21.850	AM	BROADCASTING
21.851-24.995	USB	UTILITIES (21.925-22.000 Aero; 23.200-23.250 Mil. air; 24.89-24.99 Amateur)
25.000	AM	WWV (Standard time broadcast)
25.875-26.960	USB	(Illegal CB freebanders)
26.965-27.405	AM/SSB	CITIZENS BAND
27.415-27.975	LSB	(Illegal CB freebanders)
28.000-29.700	SSB/AM/NFM	AMATEUR
29.705-30.00	NFM	LAND MOBILE
30.000-49.990	NFM	LAND MOBILE (Government and military)
50.001-53.990	NFM	AMATEUR (6 meter band) (Military)
54.000-72.000	WFM	TV CHANNELS 2-4
72.010-74.990	NFM	RELAY/REMOTE CONTROL
75.000	AM	AERONAVIGATION
75.010-75.990	NFM	RELAY/REMOTE CONTROL
76.000-88.000	WFM	TV CHANNELS 5-6
88.100-107.900	WFM	BROADCASTING
108.000-117.950	AM	AERONAVIGATION
117.950-136.975	AM	AIRCRAFT
136.980-137.980	NFM	SATELLITE
138.000-140.000	AM	MILITARY AIRCRAFT/FEDERAL
140.025-143.975	NFM	MILITARY MOBILE
144.000-148.000	NFM	AMATEUR (2 meter band)
148.025-150.775	NFM	MILITARY MOBILE
150.775-154.445	NFM	LAND MOBILE
154.45625-154.47875	NFM	POWER LINE TELEMETRY
154.490-156.255	NFM	LAND MOBILE
156.275-161.975	NFM	MARITIME/LAND MOBILE (159.810-161.550 railroads)
162.000-173.9875	NFM	LAND MOBILE/FEDERAL GOVERNMENT
174.000-216.000	WFM	TV CHANNELS 7-13
216.0125-216.9875	NFM	AMTS SIMPLEX
217.0125-217.9875	NFM	AMTS COAST
218.0125-218.9875	NFM	INTERACTIVE VIDEO
219.0125-219.9875	NFM	AMTS SHIPS
220.0025-221.9975	NFM	LAND MOBILE
222.000-224.980	NFM	AMATEUR (1.25 METER BAND)
225.000-379.975	AM	MILITARY AIRCRAFT
380-399.9875	NFM	LAND MOBILE
400.000-406.000	FM	SATELLITE/RADIOSONDE
406.1125-410.9875	NFM	FEDERAL REPEATER OUTPUT
411.0000-415.1000	NFM	FEDERAL SIMPLEX
415.1125-419.9875	NFM	FEDERAL REPEATER INPUT
420.000-450.000	NFM	AMATEUR (70 CENTIMETER BAND)

Frequency	Mode	Service
450.0125-454.6875	NFM	LAND MOBILE
454.675-454.975	NFM	AIR-GROUND TELEPHONE (GROUND)
455.000-459.6875	NFM	LAND MOBILE
459.675-459.975	NFM	AIR-GROUND TELEPHONE (PLANE)
460.000-511.9875	NFM	LAND MOBILE
512-608	WFM	TV CHANNELS
608-614	NFM	LAND MOBILE
614-698	WFM	LANDMOBILE
698-763	WFM	TV CHANNELS
763-769	NFM	LAND MOBILE
769-775	NFM	PUBLIC SAFETY
775-793	WFM	TV CHANNELS
793-799	NFM	LAND MOBILE
799-809	NFM	PUBLIC SAFETY
809-824	NFM	LAND MOBILE TRUNKED
824-849	NFM	CELLULAR MOBILE
849.-851	AM	AIR-GROUND PHONE (GROUND)
851-854	NFM	PUBLIC SAFETY
854-869	NFM	LAND MOBILE BASE TRUNKED
851.0125-868.9875	NFM	LAND MOBILE BASE TRUNKED
869.040-893.970	NFM	CELLULAR BASE
894.002-895.996	AM	AIR-GROUND TELEPHONE
896.0125-900.9875	NFM	BUSINESS MOBILE
901-902	NFM	PCS
902.0125-927.9875	NFM	CT-2/ISM/AVM/AMATEUR
928.00625-928.99375	NFM	MAS REMOTE TRANSMIT
929.0125-929.9875	NFM	VOICE AND DIGITAL PAGING
930-931	NFM	PCS
931.0125-931.9875	NFM	MAS REMOTE TRANSMIT
932.00625-934.9875	NFM	GOVT/PRIVATE MICROWAVE
935.0125-939.9875	NFM	BUSINESS BASE TRUNKING
940-941	NFM	PCS
941.00625-943.9875	NFM	GOVT/PRIVATE MICROWAVE
944.500-951.500	MFM	BROADCASTING STL
952.00625-952.9875	NFM	ALARM/PRIVATE MICROWAVE
953.000-956.050	MFM	PRIVATE/LCL GOVT MICROWAVE
956.25625-956.4375	NFM	SIGNAL AND CONTROL
956.600-959.150	WFM	PRIVATE/LCL GOVT MICROWAVE
959.85625-959.9875	NFM	PAGING
960.000-1215.000	AM	AERONAVIGATION

www.ingramcontent.com/pod-product-compliance
Lightning Source LLC
Chambersburg PA
CBHW070257230526
45470CB00002B/616